Gathering

Gathering

Using Simple
Materials
Gleaned
From The
Garden
& Nature

Linda Fry-Kenzle

krause publications

700 E. State Street • Iola, WI 54990-0001

Photography and Illustrations by Linda Fry Kenzle
Book Design by Jan Wojtech

Library of Congress Cataloging-in-Publication Data

Kenzle, Linda Fry.
     Gathering: Gleenings from the Garden/Linda Fry Kenzle.
160   cm.
Includes bibliographical references
ISBN 0-87341-557-4
Flowercrafts 2. Herbs 3. Nature Crafts 4. Gardening   I. Title.

                                                           97-80618

Acknowledgments

Appreciation is in order to all who helped me on my journey toward the garden:

Thank you to my mother for showing me the love that comes from tending flowers. Thanks to my father for his exuberant vegetable gardens.

Thank you to the Lake Geneva Public Library. When I started my journey, at age twenty, they had volumes (printed many years earlier) by Helen Moganthau Fox, Louise Beebe Wilder, Helen Van Pelt Wilson, Gertrude Jekyll, and my favorite, Rosetta E. Clarkson. What an introduction!

Thank you to Thalassa Caruso for whetting my appetite for winter gardening.

Thank you to Guri Henderson who could work magic with the bounty of nature's gifts.

Thank you to Adelma Grenier Simmons and Phyllis Schaudys for feeding my passion for herbs.

Thank you to all of the gardeners who, over the years, shared sacred information, seeds and plants.

Thank you to Deborah Faupel who instantly loved the idea of this book and shepherded it into print.

Thank you to Gabrielle Wyant-Perillo, Jan Wojtech and all of the behind-the-scenes staff at Krause who work their magic.

Thanks and love to D.C.K, J.C.K. and J.Q.K.

It was late afternoon, the music had fallen back upon the people like a heavy storm cloud which could not be dispersed to lighten and lift them, the air was growing heavy, when her eyes caught the garden as if in a secret exposure.

...[a] sudden glance seemed to have caught the garden unaware, in a dissolution of peace and greens. A light rain had washed the faces of the leaves, the knots in the tree trunks stared with aged eyes, the grass was drinking, there was a sensual humidity as if leaves, trees, and wind were in a state of caress.

Anais Nin

Ladders to Fire

To Jeremiah Kenzle
with whom I share a spiritual love of birds, butterflies and all
that is the garden.

Contents

PART FIVE: Holiday

PART SIX: In The Garden

Introduction

I wish you could be here right now to smell the delicious aromas, to see and feel the textures of the flowers, herbs and leaves. I would like you to touch the everlastings - dry but not so dry that they fall apart at the slightest breeze - like the feel of todays newspaper.

And the beautiful colors - not quite as luminous and brilliant as they were in the garden - softer, muted tones, reminiscence of old master's floral paintings. This is the elegant palette you get to work with when you create with the botanical gifts of nature.

The first section of this book offers ways to create beautiful gifts from the garden - wreaths, arrangements, baskets, a handmade journal with an accompanying treasure box and other elegant decorative pieces that will be a beautiful addition to your home whether contemporary, Early American, Country, Lodge or Antique. You'll also find beauty products made with nourishing plants - soaps, skin fresheners, a facial scrub and massage oil. Use them as gifts for friends or to refresh yourself during your personal time. To share the bounty of the garden, I've added some easy-to-make herbal foods - delicious rosemary cookies, Shrimp Seashell Pasta and a comforting herb bread.

I've presented the material on a seasonal basis - Spring, Summer, Autumn, Winter and have added a special holiday section. Each of the "Creativity" projects were selected to be appropriate for the season. **Spring** offers fresh ideas, including a twig wreath that is light and airy, representing when the first flowers comes into bloom. **Summer** is robust and outdoorsy. Here you'll find a Sea Shore Wreath and sculptural Moss and Flower Orbs wrapped in festive ribbons. **Autumn** brings a richly colored wreath done in jewel tones of garnet, peridot, yellow topaz and emerald. Graphic pods and nuts harvested in the season are nestled in a brass cachepot. **Winter** is welcomed with huge herb bundles for doors and walls. A personal Comfort Box with decoupaged pressed flowers is your gift for the season. **Holiday** is the time of glitz and glitter. A gold gilded grapevine tree illuminated to bring light to the season is a highlight. A matching gold gilded wreath is caressed in red velvet ribbons.

At the end of each seasonal chapter you'll find **The Pleasures of the Season**, a gathering of ideas and things to do in the garden. This collection is simply for fun. Take a nature walk or learn how to attract birds and butterflies to your garden and make a flower press. Make nature a part of your everyday lifestyle enjoyments.

If you enjoy working with the fragrances of the flowers, please savor the last section titled **In The Garden**. You can grow your own everlastings and herbs. Use the charts to plan a glorious garden. Specific plants listed in special charts attract birds, hummingbirds and butterflies. Welcome some of these elusive flyers into your garden landscape.

The pleasures of gardening are many. Fresh air, beauty and creativity. These gifts are yours in the garden. Escape to your own backyard sanctuary. Enjoy the spiritual caress of nature. Embrace the scents, textures and the promise of annual renewal.

L.F.K.
Little Eden
1998

"Live in each season as it passes;
breath the air, drink the drink,
taste the fruit,
and resign yourself
to the influences
of each."

-Henry David Thoreau

PART ONE:

Spring

Spring is the season when the ground begins to thaw, the soft rain showers come, and everything in the garden begins to awaken. Leaf buds start to unfurl and bulbs planted last fall are now sprouting from the ground. Tulips and daffodils join the party a little later in the month and the robins return. Spring is a time of awakening, a time of renewal when everything is once again possible.

Lexicon Botanical Basket

To get the refreshing scent of green in your house, a few weeks before the Easter Holiday plant a hand-decorated basket with grass seed, chives or alfalfa. I used an oval gathering basket decoupaged with botanical illustrations from seed catalogs. On the top and bottom basket rings I wrote a continuous message, "A Garden's Gifts are Many." This makes a playful holiday centerpiece. After the holiday, remove the soil and use the basket for gathering the gifts of the garden.

CREATIVITY

1. Paint the center band of the basket and let dry. I used a dark forest green paint. Any shade of green or a sky blue would look good.

2. Look through catalogs and magazines for botanical illustrations to fit the space on the band. I chose spring bulbs - tulips, daffodils, early lilies, irises and snow drops. Cut out fifteen illustrations and place them around the center band. When you have a nice arrangement, coat the back of an illustration with the liquid podge and adhere it to the center band. Attach all illustrations and let dry. Apply a coat of liquid podge over the center band.

3. Write a repetitive message on the top and bottom band.

4. Line the basket with plastic and trim as necessary.

5. Fit the basket with potting soil.

6. Plant seeds by scattering them on the surface of the soil. Water with a mist sprayer and cover the top of the basket with plastic wrap to encourage germination.

7. When the seeds sprout, remove the plastic wrap.

8. Add colored Easter eggs.

POSSIBILITIES

🌱 Make a huge herb garden basket. Line a large handled basket with plastic (add drainage holes) and fill with soil and plants. Place the basket on a deck or near the kitchen door.

🌱 Using the same concept as above, plant a group of smaller baskets with ivies of different leaf shapes.

🌱 Pansies are available in the spring. Plant a strawberry basket with pansies and place it by the door.

🌱 Hang a pocket basket on an outside wall near the door. Fill it with soil and grass seed and attach a banner that says, "Happy Spring!"

🌱 Easter place "cards": Place soil in an eggshell and plant with grass seed. Write a name on the eggshell.

"We are stardust, we are golden, and we've got to get ourselves back to the garden."
-Joni Mitchell

Nature Print Journal

Construct your own journal to hold private thoughts, addresses, anniversaries and birthdays or whatever you need to inscribe for perpetuity. This journal is made of fine French paper embellished with delightful prints of leaves.

One sheet Arches watercolor paper
Twelve sheets 11" x 7" vellum paper, white
One sheet 11" x 17" vellum paper, pink or
    other contrasting color
Delta™ Shimmering Fabric Colors: hunter
    green, leaf green, sachet, denim blue,
    red, champagne and coral (or other
    colors of choice)
Paint brush,1/2" flat
Paper plate or other palette
Glass of water
Paper towels
Fresh leaves
Five small star-shaped studs
One yard satin twist cord,
    1/4" in diameter
Steel ruler
Rotary cutter
Sea sponge
Rubber cement
Paper punch

CREATIVITY

1. Use the rotary cutter and ruler and cut two pieces of watercolor paper 11" x 17". On the first sheet of paper, paint the short edge of the paper with the sponge painting technique. Squirt a small amount of each of the green paints and a bit of coral and champagne. Dip the sea sponge into the paint and press it onto the paper. Continue sponge painting until a 2" section is fully covered. Repeat on the other short edge of the paper.

2. Fold the other piece of watercolor paper in half. Use a rotary cutter and cut 2 1/8" off the

short ends. The paper should now measure 11" x 12 1/2". Paint one of the 1" x 11" cut off pieces of solid denim blue and let dry. Rubber cement this piece to the plain piece of watercolor paper so only 1" of the blue paper is exposed.

3. Gather leaves. Coat each leaf with paint and gently press it onto the watercolor paper. Try using two or three colors on one leaf. Experiment printing each leaf more than once. Continue printing leaves until the cover section of the watercolor paper is covered. Let the paper dry. If you wish to create the leaf box in the next companion project, keep each painted leaf after it has been printed.

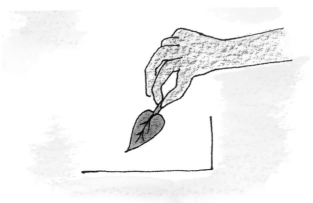

4. Rubber cement the two pieces of painted watercolor paper together. Weigh down if necessary.

5. Fold all of the white vellum paper in half. Fold the pink paper in half. Open up the sheets and lay them in a stack. Lay the cover on top of them. On the spine fold, measure and mark 1" up from the bottom and 1" down from the top. Make the holes using the paper punch.. Mark and punch the rest of the paper.

6. Feed the cord through the holes and tie in a knot. (See illustration on page 22.)

7. Use the rotary cutter and ruler to cut away paper extending beyond the edge of the book.

8. Glue star-shaped studs onto the blue strip.

9. Optional: If you wish to use the leaves for the following box project, allow the paint on the leaves to dry and place the leaves between the pages of a telephone book or in a flower press.

POSSIBILITIES

🍃 Bind the book using a long-throated stapler.

🍃 Simplify the book by constructing it of two 11" x 17" sheets of water color paper and rubber cement together.

🍃 Make the entire journal out of water-color paper and use it

for water media drawings.

🍃 Any acrylic paint will work for nature printing.

🍃 Tie the journal with ribbons.

🍃 For an address book, attach index tabs to the paper edges.

"...flexibility, an openness to adventure, and a willingness to experiment, play, take imaginative chances, and follow your path are essential in the [journal-making] process."
-Tristine Rainer

Leaf Box

Every woman needs a special grace notes box on her dresser. Each night, on a slip of paper, write down something for which you are thankful. Fold the paper up and put in the box. While doing this, light the star-studded tapers. This is a beautiful expression of love and it will allow you to see the bounty of your life. Of course, you can also use this beautiful box for jewelry, mementos or other precious artifacts.

Wooden box
Painted leaves from journal project
Galleria™ acrylic paint: blue, white
Paintbrush
Palette
Paper towel
Masking tape
Liquid podge
White tacky glue
Optional: two white tapers and star-
   shaped studs

CREATIVITY

1. To create a frame effect, use masking tape to mask off a border around the top of the box.

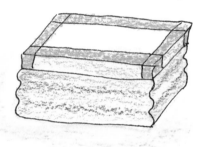

2. Paint the top of the box white and let dry.

3. Remove the masking tape. Paint the border blue and let dry.

4. Squeeze out a large dab of blue and white paint. Mix the paint together to create a graduated palette ranging from dark blue to medium blue to very pale blue. Paint the sides of the box. Use the most pale blue on the bottom and work to the darker blue at the top. To prevent the lid from sealing itself closed, allow the paint to dry a bit and place something in the box to hold it open. Let the paint dry thoroughly.

5. Coat the top of the box with liquid podge. Begin arranging the dried, painted leaves in place. Move the leaves around to find a pleasant arrangement. Add another coat of liquid podge on top of the leaves and let dry.

6. Coat the rest of the box with liquid podge and let dry.

7. Push the spikes of the star-shaped studs under so the stars will lie flat. Glue the studs on the box.

8. Optional: Insert the star-shaped studs into the tapers.

POSSIBILITIES

🌿 Transform any box - a cigar box, an old wooden jewelry box or a heavy cardboard box.

🌿 Use this box to hold photographs or your Mother's Day and birthday cards.

❦ Stamp the box with rubber stamps for an easy decorative effect.

❦ Glue thick dimensional dried flowers to the box lid, then coat with acrylic gel to seal.

❦ Use this technique to create a set of old-fashioned canisters for your kitchen or containers for your bath.

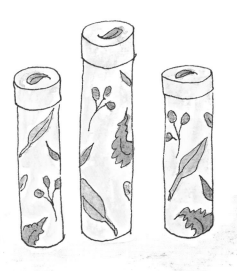

"Long before women had claimed their present independence in the arts and professions, in trade, in sport, in travel, and in many difficult crafts, she had quietly and firmly established her right to self-expression."
-Lady Jekyll on her sister-in-law Gertrude Jekyll

Twig Wreath

Before the trees and shrubs begin to leaf, gather a bunch of twigs to create a wispy twig wreath. This ethereal wreath is small enough to brighten any little wall space. Add a nice selection of dried flowers as a reminder of the coming lush garden growth.

Twigs from shrubs or trees cut to a
  6" length
A 12" wire wreath frame
Florist's wire
Purple and yellow statice
Love-in-a-mist
Glue gun

CREATIVITY

1. Gather the twigs into small bunches and wire at the base to hold.

2. Wrap the twig bunches onto the wire wreath frame with florist's wire. Continue working around the frame, covering up the wire of the previous bunch with the current bunch. Fill the entire wreath frame.

3. Cut all of the dried flowers to a 3" length.

4. Use the glue gun to attach each flower to the wreath. Hold the flowers in place for a moment to be certain that the flowers remain erect. Strive to achieve an aesthetic scattering of colors and textures.

5. Wire a hanger to the back of the wreath.

POSSIBILITIES

❦ Glue kitten-soft pussy willow catkins all over a twig wreath.

❦ Make tiny twig wreaths for hanging in windows and, if you like, attach tiny birds.

🌱 Spray the wreath gold or silver and attach tiny Christmas balls to display later in the year.

🌱 Make the wreath to a theme: For a fisherman, attach interesting lures; for a baker, attach small cookie cutters; for a seamstress, attach sewing notions; for a carpenter, attach children's tools. Pick your own theme!

"I came to understand that I was happiest in the sunshine, playing with my flowers, fruits, vegetables, inspired by the colors and patterns, the smells and the beauty of nature."
-Alexandra Stoddard

Rosemary Cookies

The distinct flavor of these cookies will get you ready for all of the mouth-watering garden tastes and aromas to come. You must use fresh rosemary in this recipe. If your garden isn't producing yet, rosemary is available in most markets. Enjoy these crispy cookies with a cup of tea and a good book.

NECESSITIES

1/2 cup butter
1/4 cup sugar
1 egg
1 1/2 cup flour
1 teaspoon fresh rosemary leaves
Bowl
Rolling pin
Wooden ladle
Heart-shaped cookie cutter
Cookie sheet

CREATIVITY

Preheat oven to 375 degrees.

1. Microwave butter on HIGH for 1 minute. Stir in sugar and rosemary. Add egg. Fold in flour. Mix thoroughly.

2. On a well-floured work surface, roll out dough to about 1/4" thick. Cut with the cookie cutter and place hearts on a cookie sheet. **Bake for 8 to 10 minutes.**

Makes approximately 27 small heart-shaped cookies.

POSSIBILITIES

♡ For a tart cookie, try lemon verbena leaves and a dash of lemon zest.

Lemon verbena

♡ For an exotic flavor, use crushed coriander.

♡ Lavender makes a delicately flavored cookie.

lavender

♡ Place any plain cookie in a tin with dried rose petals. After two weeks the cookies will have taken on the rose scent. (Try this with sugar too.)

♡ Use ready-made frosting to add a candy message heart to the center of each cookie. Happy Valentine's Day!

Give a little basket of cookies wrapped in pink tissue paper to a friend or neighbor that needs a little boost to get through the final phase of the cold weather.

Cling, swing, spring, sing, swing up into the apple tree.
-Keats

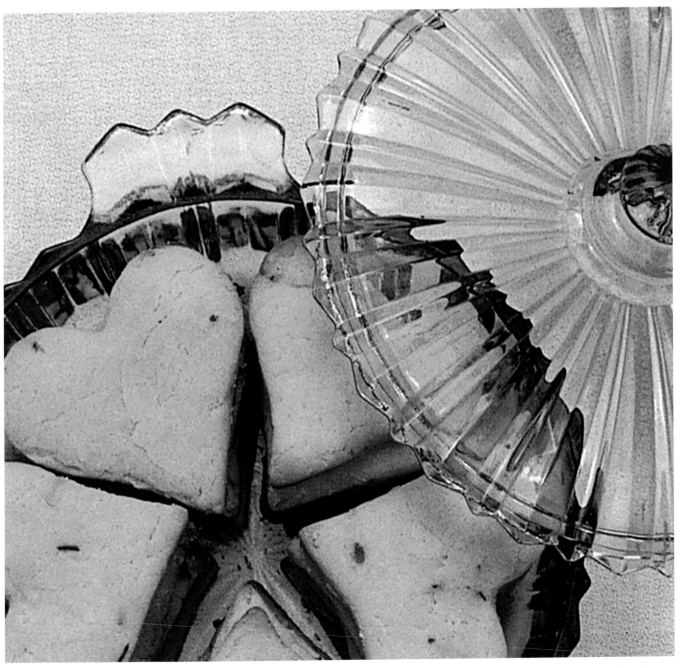

Lavender Herb Soap

Take a moment to refresh yourself in a luxurious lavender herbal soap bath. Wash your worries away and smell delicious too! Place these scented soaps in a gift basket and consider giving them to a friend who needs a gesture of love.

NECESSITIES

Three bars Ivory™ Soap
1/2 cup water
1/3 cup lavender flowers
Gift basket
Small white basket
Twisted craft paper, white
Lavender tissue paper
Two tiny roses
Florist wire

CREATIVITY

1. Grate the three soap bars. Place the flakes in a bowl.

French Lavender

2. In a small microwaveable bowl add the water and lavender flowers. Microwave on high for one minute.

3. Pour the liquid onto the soap flakes and knead into a loaf. Separate into three equal size pieces and quickly roll into balls. Set balls aside to dry.

GIFT BASKET

1. Cut six pieces of tissue paper to fit into basket.

2. Cut two lengths of twisted craft paper, unwind and fashion into bows. Cut two pieces of wire. Insert a wire through the knot at the back of the bow. Attach the bow to the basket handle. Insert a tiny stemmed rose between the bow and the basket handle. Repeat for the other side of the handle.

3. Add lavender soap balls and a tiny note card.

POSSIBILITIES

🌱 Try thyme, anise hyssop, pineapple sage, basil, roses, rosemary or patchouli for a different scent.

🌱 Use Dove™ soap as a base for a creamier, less drying soap.

🌱 Shape the soaps using cookie cutters or designer ice cube trays.

🌱 Remove 1/2 teaspoon scented water and add 1/2 teaspoon aloe for a soothing, healing soap.

🌱 To compliment the herbs, wrap each soap in tissue paper scented with one drop essential oil.

> "I go to Nature to be soothed and healed, and to have my senses put in tune once more."
> -John Burroughs

THE PLEASURES OF
Spring

Gather branches for an early touch of spring. Collect forsythia and pussy willow when the buds are just starting to break. Force them by plunging the branches into a large vase filled with warm water. Change the water daily and soon the branches will bloom. As an added bonus some of the branches will root in the water. Plant them outside as soon as the soil has warmed enough to be entrenched. For the first year, water for five minutes once a week. Note: This is a good technique to establish any tree, shrub or perennial.

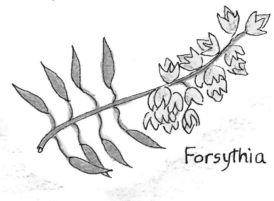

Forsythia

Start collecting flowers. Take tulip petals and daffodil florets, just before they fall from the stem and allow them to dry on a sheet of thick paper toweling.

Create lavender flower dryer sachets to add a delicious scent to your laundry. (See Sachets on page 75.)

Sweet Lavender

Watch your local newspaper for news of proposed housing developments. Contact the builders and ask permission to dig up the native plants. Create a Wildflower Rescue Garden

using Richard Bir's Growing & Propagating Native Wildflowers as an excellent reference guide.

Make a flower press: Cut two pieces of 1" x 6" lumber into 10" pieces. Drill a hole 1" in from each of the corners. Purchase four bolts, eight washers and four wing nuts. Set aside. Cut newspaper into 8" squares. Use the newspaper between the fresh flowers to absorb the moisture. Place the bolts and washers through the holes of one of the wooden squares. Add the stack of newspaper and fresh flower petals. Place the second wooden square in place on the bolts. Add another set of washers and tighten down the wing nuts. Most flowers take about two weeks to dry. When totally dry remove the petals and store them in envelopes.

a.

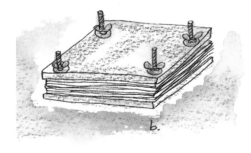

b.

Use some of your pressed flowers to make a charming lamp shade decoration. Select a smooth shade without any texture. Adhere flowers and leaves in a pleasing arrangement with liquid podge and let dry. To seal, coat the entire shade again with liquid podge and let dry.

No green thumb? Plant shrub roses. They are very hardy, tolerate most soil conditions and many are pleas-antly

scented. As a bonus, when the roses have finished blooming in the fall, you will be left with rose hips. These are the type of roses I use in foods and moist potpourris.

Plant a theme garden - an herb garden, a butterfly garden or a bird-attracting garden. See the appendices for detailed information as to plant choices and additional informa-tion.

Add charming trellises and arbors to your garden landscape. Plant them with flower-clustered wisteria, chocolate-scented Akebia quinata, kolomikta vine with pink and white splashed green leaves, Virginia creeper, fragrant honeysuckle, huge white moonflowers or beautiful blue morning glories.

Plant hanging baskets with blooming orange-eyed black-eyed Susan vine, lush green ivies or ivy-leaved geraniums. These bloom in a riot of colors ranging from flaming scarlet to soft apricot and innocent pinks and whites.

Divide perennials as soon as the soil can be worked. Perennials will revive faster at this time of year. Later in the year it may be very hot and stressful.

Make a butterfly box. This is the plan for a butterfly box similar to the cedar one that my husband made for me. The front is assembled so that two long open gaps allow the butterflies to enter. Place a stick or two in the box so the butterflies can make a chrysalis.

Cut out wooden bird shapes such as distinctive cardinals. Paint and seal the shapes with varnish and mount them on your garden shed.

Think Christmas! Plant evergreens at the back of your property that can be harvested in years to come and used as Christmas trees.

Don't forget

picturesque window boxes. Plant them with bloom-crazy annuals for brilliant color all summer long. Add a vine or two for drama.

Gladiolus

🌱 Start planting gladiolus bulbs at about fourteen day intervals for a riot of colorful spires in the summer.

🌱 Pick a handful of lily of the valley by pulling on the stems. They slip right out without needing to be cut or snapped. To capture their dreamy scent, display the dainty flowers in small vases all over your house.

🌱 For a rapid bust of new growth, cut off the first rose buds down to the first five-leaflets.

🌱 Garden easy. Plant any no-traffic areas with

myrtle

ground covers. It looks delightful and no mowing is required! Try crown vetch, lily of the valley, variegated goutweed, bronze bugle and blue-flowered myrtle.

dandelion

🌱 Dandelions are popping up all over. Show a child how to make a dandelion necklace by tying the stems together. Place a dandelion under a child's chin and show them how the glorious sunshine reflects into their spirit.

🌱 Let a child make his/her own garden. Scrape off a bit of lawn and encircle the space with bricks, rocks or a ready-made wood edging. Let the child pick out a flat of colorful annuals and show him/her how to plant seeds of sunflowers, green beans and nasturtiums.

🌱 Pick an armload of lilacs. Strip away the bottom leaves, fill a huge vase and place them

in a central room in your house. Delicious scent! Whisper thank you for the bounty.

🌱 Hang bird houses. Note: Bell your cat.

🌱 Start a garden journal. Pick out a large hardbound blank page book. From garden catalogs cut out pictures of plants that were ordered in the spring. Cut out pictures of beautiful gardens, articles and tips from garden magazines. Glue these into your journal. Add labels from locally purchased plants. Add pressed flowers. Collect garden quotes or start a bird list. Write down bloom times, how a plant performed and if you liked it. Design a bird bath or a garden sculpture. Make sketches of garden plans, paint plant portraits or keep track of the weather. Talk to your soul.

PART TWO:
Summer

The hot breath of summer warms our flesh and the garden delights in the heat and humidity. Annuals bloom like crazy and perennials grow strong and hardy. Summer is a busy time of picnics with friends, pool parties and nature hikes. All of the summer birds are now in their birdhouses raising broods. Spend time on your patio or deck relaxing and enjoying the existence of nature all around. Listen to the songs of summer.

Birdhouse Basket

I love wild birds in the garden. Special baskets like the one shown are available at craft centers. Or, if you prefer, choose a plain basket and work up a graduated flower design.

NECESSITIES

Basket
Oasis or Styrofoam
Dried flowers (I used pepperberries, statice and other small flowers)
Spanish moss
Wired bird
Wire cutters

CREATIVITY

1. Insert the Oasis or Styrofoam into the basket. Cover with a layer of Spanish moss.

2. Begin the flower design at the front of the basket. Insert pepperberries. Add row upon row of flowers by trimming the stems to the same length.

3. With wire, attach the bird to the handle. Trim away excess wire.

POSSIBILITIES

🌱 Use brightly colored flowers at the front of the basket and graduate to darker colors at the back. This gives the illusion of depth and the sense of a grand garden.

🌱 Fill the basket with just one kind of flower that you love.

🌱 Transform the basket into an outdoor bird feeder. Spread suet on the house and sprinkle with bird seed. Add orange slices in the basket.

🌿 Do the basket all in ivy and twine a stem or two around the basket handle.

🌿 Use a basket without a handle. Tuck a square of Styrofoam in the basket and add tall candles. Add dried flowers in a low-slung arrangement.

There is a true music of Nature;
the song of the birds,
the whisper of leaves,
the ripple of waters upon a
sandy shore,
and the wail of wind or sea.
-John Lubbock

Fragrant Rose Wreath

One of the pleasures of summer is the alluring aromas of roses drifting through the air. I love to sink my nose into the heady blossoms. This wreath has eighteen roses of three different colors. You must pick the roses before they start to unfold or they will fall apart when dried. Pick the buds and hang them to dry. The base is made of salal leaves (lemon leaves) with accents of pink pepperberries and wisps of German statice and babies breath. Beautiful simplicity.

NECESSITIES

Eighteen roses (I used six deep red, six
   white that dried a beautiful
   champagne, and six golden yellow
   tipped with scarlet red.)
Salal leaves
14" wire wreath frame
Florist's wire
Pepperberries
German statice
Babies breath
Glue gun

CREATIVITY

1. Gather the
salal leaves into
bunches and wire
them to hold.

2. Overlapping them as you go, attach the
leaf bunches to the wreath frame. Fill the entire
frame.

3. Between the leaves, glue in six of the
roses. Space them an equal distance apart.
Glue in the next color of roses then add the
final six roses.

4. Fill in with bunches of pepperberries.

5. Add wisps of German statice and babies
breath to soften the design.

6. Add a wire hanger to the back of the
wreath.

POSSIBILITIES

Scent the wreath with drops of attar of
roses.

For a rich, gilt look, use a small artist's paintbrush and gold enamel paint to brush the edges
of the roses.

Add sprigs of bright green Sweet Annie to
add a depth to the greens in this wreath. You'll
also get a powerful aroma!

🌱 Choose groupings of other favorite flowers for this wreath.

🌱 Make it Victorian by adding skinny satin ribbon. Weaving the ribbon through the flowers and leaves of the wreath.

Moss & Flower Orbs

I love the sculptural quality of these balls. Place them on a marble topped table for a stunning contrast. Beautiful.

NECESSITIES

Styrofoam balls: 8" and 6"
Moss
Pins
Ribbon, 1/4" wide, 3 yards per ball
Flower potpourri
White tacky glue

CREATIVITY

1. Open up moss into flat sheets. Lay moss on a ball. Wind the end of the ribbon on a pin and insert into the ball. Begin wrapping moss and add extra moss when necessary. When the ball is completely covered with moss, wrap ribbon around and around. Wind the ribbon end around another pin and insert into the moss ball.

2. For a potpourri ball, apply glue on half of a Styrofoam ball. Sprinkle with potpourri and let dry. Repeat for the other side of the ball.

3. Wind the end of ribbon on a pin and insert into the ball. Wrap the ball. To secure the end, wind the ribbon end on a pin and insert into the ball.

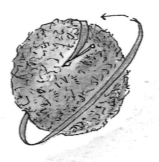

POSSIBILITIES

🌱 Make a kissing ball by adding mistletoe. Suspend the ball in a doorway.

🌱 Attach strands of fresh bittersweet with fern pins.

🌱 Glue the Styrofoam ball into a clay pot. Cover with moss or flower potpourri.

🌱 Use the ribbon to create a floppy bow and pin it to the top of the ball.

🌱 For a very textural piece, cover the balls with rose hips or globe amaranth.

🌱 Glue seashells to the ball. Use one kind or add many types for variety.

" What is a weed? A plant whose virtues have not been discovered."
-Emerson

Seashore Wreath

It is so much fun to discover sea shells peeking out of the sand. To remind you of lazy days at the beach, collect a full bag of shells and create this lovely wreath.

Straw wreath form, 12" in diameter
Sea shells
Spanish moss
GOOP™ adhesive
Wire for hanger

CREATIVITY

1. Attach a piece of wire around the wreath form to create a loop for hanging.

2. Glue shells in a pleasing arrangement on the face of the wreath. To create an interesting arrangement, remember to overlap the shells and pile some of the smaller shells upon the larger shells. Let glue dry.

3. Wrap the exposed areas of the wreath with Spanish moss. Hang.

POSSIBILITIES

♥ Add a bit of fishing net and a large fishing lure. Give this wreath to your sweetheart.

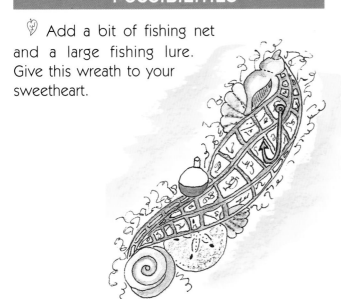

♥ Glue a mirror behind the wreath.

♥ Tie on tiny bottles filled with secret messages.

♥ Use tiny wreaths covered with seashells as candleholders.

♥ Fill interesting clear bottles with seashells; add a votive candle. Place them all over your house.

"Nature will deliberately reveal itself...if only we look."

-Thomas Edison

Seashell Pasta

Fresh shrimp mingle in a blanket of seashell pasta, scented with fresh herbs in a light vinaigrette. Lovely for a late summer supper.

Shell pasta, 7 oz.
1/2 pound small shrimp
1/4 cup green pepper, diced
1 carrot, shaved
1 clove garlic

Vinaigrette
1/2 cup olive oil
1/4 cup vinegar
1/4 slice lemon
1 tablespoon parsley
1/2 teaspoon dill
1/4 teaspoon marjoram
1/4 teaspoon thyme

Saucepan
Large bowl
Knife
Blender

CREATIVITY

1. Cook pasta following directions on the package. Drain.

2. Mix together pasta, shrimp, green pepper, carrot and garlic.

3. Use a blender to mix oil and vinegar. Grate and add lemon zest. Squeeze lemon quarter into blender. Add herbs. Blend and pour over salad. Refrigerate to blend flavors. Serve with additional slices of lemon.

POSSIBILITIES

Add a 1/4 teaspoon of Dijon mustard to vinaigrette for a spicier taste.

Add a few shakes of Tabasco to the final salad. Mix thoroughly before serving.

If price is a factor, or just for a different taste, try surimi.

Try other pasta shapes such as farfalli, mostaccioli, vermicelli, etc.

Sprinkle chopped black olives over the salad.

"If an event is meant to matter emotionally, symbolically, or mystically, food will be close at hand to sanctify and bind it."

-Diane Ackerman

Lavender & Flower Candles

Candles are lovely mood enhancements any time of the year. Use them on a summer porch, on a late summer afternoon while watching fireflies, or in the bath.

Lavender Candle
One tall pillar candle, white
Lavender flowers
Sheet of newspaper
One long cinnamon stick
Raffia
Dried flowers or pods
Liquid podge
Paintbrush, 1" flat

Flower Candle
One squat candle, champagne
Pressed flowers and leaves
Liquid podge
Paint brush, 1" flat

CREATIVITY

Lavender Candle

1. Pour lavender flowers onto a sheet of newspaper.

2. Coat sides of the candle with liquid podge. Immediately roll the candle in lavender flowers and let dry. If there are any bare spots, add more liquid podge with a paintbrush and press on additional lavender flowers. Let liquid podge dry.

3. Tie cinnamon stick in place with raffia. Add additional dried flowers and pods as desired.

Flower Candle

1. Coat the sides of the candle with liquid podge. Arrange the flowers and leaves. To seal in the botanicas, add an additional coat of liquid podge over the candle sides.

POSSIBILITIES

Dried flowers can be attached with a small artist's paint-brush and egg white.

Roll a candle, coated with white glue, in crushed rose petals or any other dried petals left over from other projects.

Roll a candle in seeds. For example, garden seeds, poppy seeds or coriander.

Wrap the candle in beautiful ribbons.

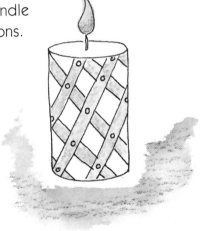

"The earth can be an abundant mother if we learn to use her skill and wisdom-to tend to her wounds, replenish her vitality, and utilize her potentialities."

-John F. Kennedy

THE PLEASURES OF
Summer

🌿 Attract hummingbirds to your garden by planting deep-throated, nectar-rich flowers in hot colors (reds, yellows, oranges, and pinks). Good perennial choices are: Columbine, honeysuckle, trumpet vine, phlox and monarda (bee balm). For the annual garden try petunias and impatiens. All of these flowers should be planted in a sheltered spot away from high winds.

🌿 Hang a red hummingbird feeder filled with sugar water. Boil one quart water and stir in two cups sugar. Continue stirring until sugar is completely dissolved. Do not add red food coloring; it is better and safer to use a colored feeder. Allow the sugar water to cool. Fill feeder and hang. Store the extra sugar water in a clean jar. Every two weeks clean the feeder and replace the sugar water.

🌿 If you notice a hummingbird that doesn't quite look right, you probably have attracted a hummingbird moth. Look it up in your field guide.

🌿 Lay a string of clear Italian lights in the bottom of a large fish bowl. Add a layer of seashells. Cover the cord and each light with a shell. Place an additional shell over each light on the cord that comes out of the fish bowl. Plug it in. Beautiful. Use it as a centerpiece on a sideboard or simply set it in the corner of your living room. The glowing bowl sets a gracious mood.

🌿 Edible flowers: Plant some edible flowers to decorate a salad or cold pasta. Squash blossoms, nasturtiums, violets, calendula flowers are all good possibilities. All flowers are not edible. Some will make you ill; others are potentially fatal. Check the bibliography for books on edible flowers. Then select and plant a little edible flower garden. Add herbs for variety. Note: Herb leaves and flowers are edible.

🌿 Candied violets: Use gracious candied violets on cakes and tea breads. Coat the front and back of each flower with egg white. Spoon granulated sugar over the flower. Let flowers dry on a cookie sheet. These can be stored for one year.

🌿 For a basic herb garden and your cooking pleasure, plant these five herbs: parsley, sage, rosemary, thyme and basil. Consult the herb garden chart in the reference section for information on these five herbs and many more.

sage

Save, in paper bags, all of the cut-offs and stems from your herb garden. For a beautiful scent, sprinkle the gleanings into your fireplace. Give bags of dried herbs to friends. They will enjoy the dreamy scents.

Make lavender bottles: Take a small handful of fresh lavender and tie it together right under the bottom florets. Bend down the long stems so the florets are inside a basket of sorts. Tie the stems together at the bottom and tie on a ribbon. Decorate with a few dried flowers. Use the lavender bottles as linen fresheners, present tie-ons or ornaments.

relax in the garden looking at the beauty you have created.

Make a simple plaster bird bath while at the beach. Take along a box of plaster of Paris, a large bucket and a stirring stick. Carve a shallow hole in the sand. Mix up the plaster according to directions on the box. Put the plaster into the recess in the sand. Press shells, beach glass and other interesting things into the plaster. Let the plaster dry. Take the new bird bath home at the end of the day.

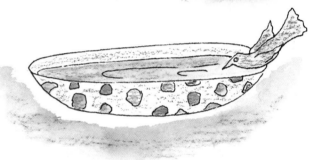

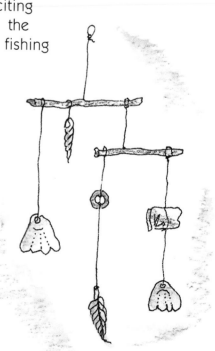

Use a collection of shells and driftwood to make an exciting mobile. Tie on the pieces using fishing line, find the balance spot and tie off. Continue adding as many shells as desired. Hang it on a porch or deck.

a.   b.   c.

Place a few sprigs of lavender or other cooking herbs in a bottle of white wine - Vouvray or Cabernet Sauvignon are good choices. Replace the cork and let the bottle stand for at least two weeks. Then enjoy as you

The sound of water is so soothing that you may wish to incorporate it into your garden. Place a submersible pump in a wine half-barrel. Add a few water plants elevated by bricks. Turn on the pump and enjoy.

Make your own stepping stones: You will need a bag of Sakrete, a large bucket, aluminum roasting pans, pieces of window screening for reinforcement and water. Mix up the Sakrete following the instructions on the bag. Pour about 1" of concrete in one of the roasting pans, add a piece of screening, then add another 1/2" of concrete. Before it sets, add bits of broken tiles or plates, rocks and sand. Or, wait until it is nearly set and use a stick or dowel to add a message: "Welcome to my Garden," "Relax," "Enjoy," or some other delicious intent.

PART THREE:
Autumn

The year starts to slow down. The summer birds head south. The brilliant colors of summer turn to the deep, rich colors of autumn - golden yellow, russet, scarlet, terra cotta and mahogany. The sun is lower in the sky. It is the time to harvest the bounty of the gifts of the garden and field. It is a time of reflection.

Maple Leaf Wreath

Maple leaves have the coloration of fine garnets or a nicely aged red wine. Such a beautiful red color. The maple leaves are accented with green eucalyptus, bunches of wheat, dried black-eyed Susans and stems of naturalistically styled grapes. This is a nice color grouping and the eucalyptus emits a delightfully fresh scent for a long time.

NECESSITIES

An armload of glycernized maple leaves
Eucalyptus
Wheat
Black-eyed Susans
Two bunches of grapes
Two preserved slices of orange
Purple statice
Tansy
Serum flower heads
Wreath form, 16"
Florist wire
Glue gun
Raffia

CREATIVITY

1. Using wire, attach stems of maple leaves to the wreath form. Strive to get an even amount all around the circle. Glue in any stray leaves as necessary.

2. Glue the sprays of eucalyptus to the center bottom of the wreath. Add a tuft of wheat, and glue on the two preserved slices of orange. Glue on a few stems of purple statice on either side of the orange.

3. Using wire, attach the two bunches of grapes. Tie with raffia for a natural appearance.

4. Add additional flowers and wheat stems into the rest of the wreath. Create a pleasing arrangement. If you cannot find the flowers suggested, use others for a different effect.

5. To the back of the wreath form, create a wire loop for hanging.

POSSIBILITIES

🍃 Use the same necessities to create a topiary instead of a wreath. Build it up on a Styrofoam cone form.

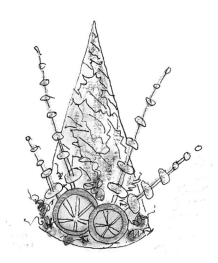

🍃 For another autumn look, use fragrant eucalyptus instead of maple leaves, silvery Artemisia instead of wheat and huge golden grapefruit slices instead of the oranges. Beautiful and very fragrant too.

🍃 Gather dried grasses on your walks and incorporate them into a wreath.

🍃 Dry some colorful foliage, tuck the leaves in the wreath here and there, to add touches of gold, orange and green.

🦉 Use a bundle of dried corn husks instead of the wheat. Add a cluster of strawberry popcorn or Indian corn.

"The grim frost is at hand, when apples will fall thick, almost thunderous, on the hardened earth.

-D.H.Lawrence

Harvest Basket

Gather and dry the gleanings from your garden and the gardens of friends and relatives. If you find beautiful items at your local flower shop, and they make you smile, add them too. The basket shown includes velvety red cockscomb heads, huge hydrangeas in sophisticated muted colors, bright white strawflowers, salal leaves, spires of lavender, preserved autumn leaves, magenta rattail statice, shocking pink pepperberries, bright purple statice, love-in-a-mist, lemon yellow carnations, pink yarrow and orange Chinese lanterns. The eucalyptus and dried oranges are signature elements and should be added to any harvest basket collection.

Natural grapevine basket with a handle
Eucalyptus
Two dried oranges
A diverse collection of beautiful dried
  flowers
Styrofoam
Moss
Florist's wire (optional)
Sticks or dowels (optional)

CREATIVITY

1. Attach any flowers with wire, like the cockscomb, by adding a plant stick or dowel. Attach the strawflowers by inserting a wire into the base of the flower head. Cut the stemmed flowers to a manageable length of 10".

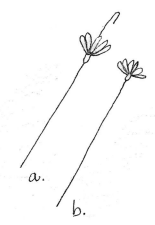

a.

b.

2. Weave lengths of eucalyptus around the basket handle. If necessary, secure with florists wire.

3. Insert the Styrofoam into the basket.

4. Begin inserting the stems of long flowers (rattail statice, lavender, salal leaves) in the center of the basket. Continue adding groups of flowers until you reach the basket edge. Study

the arrangement. If you love it, leave it. Otherwise, carefully remove a flower bunch and replace it with something else. The flowers are more sturdy than they appear and can tolerate some manipulation.

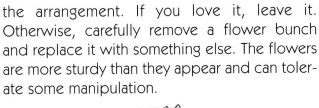

POSSIBILITIES

When the berries of orange bittersweet are ripe, before they open, gather stems. Wrap the fresh branches around the handle and the bottom of the basket. When the bittersweet berries open they are a delicious orange color. Fill the basket with dried autumn leaves in opulent golds, bright yellows, striking tangerines, scarlet reds, rich mahogany and charming greens. You may wish to use this as a burden basket. Write your worries on slips of paper and tuck them down in the leaves. Leave them for a different day when you feel strong.

Spray the basket white and fill it with wedding flowers.

Tie the flowers into beribboned bundles and hand them out to visitors.

"If you take any flower you please and look it carefully over and turn it about, and smell it and feel it and try to find out all its little secrets...you will discover many wonderful things."

-Gertrude Jekyll

Nuts & Pods in Brass

Some of the most intriguing textures are found in nuts, pods and fungi. The next time you are out on a nature walk, look on fallen trees for shelf lichens and other interesting fungi. These feel dry and can be removed from the tree with a pocket knife. Most pods are seedheads from trees so keep an eye on the ground for these treasures. If you live in a city, make a pleasant arrangement out of store-bought nuts in the shell - walnuts, hickories, pecans, etc. Craft stores also carry a selection of interesting items such as lotus pods. After you have collected and dried the nuts, pods and fungi, place them in a plastic bag and scent them with fifteen drops of vanilla essential oil. Keep the bag closed for two weeks. Dry strips of orange peel by simply placing them on a paper towel. Select and use a container that you love. I used a warm brass cachepot to compliment the brown colors.

NECESSITIES

A collection of nuts, pods and fungi
Two long cinnamon sticks
Oasis or Styrofoam
Moss
Container of choice
White glue
Vanilla essential oil

CREATIVITY

1. Place Oasis or Styrofoam into a container and cover with moss.

2. Implant the cinnamon sticks to one side of the container.

3. Arrange the fungi and lichens around the edge of the container and glue in place. Begin building up the design by contrasting shapes. For instance, a round nut next to a conical pod. When a desirable arrangement is complete, glue in place.

4. Tuck in dried strips of orange peel.

POSSIBILITIES

For a novel presentation, glue nuts and pods to inexpensive glass candlesticks.

Attach nuts and pods to a small open-top box. Fill with row upon row of long cinnamon sticks.

Capture the essence of a forest: Glue nuts and pods to the edge of a large mirror. Every time you look at your reflection you'll be transported to a north woods lodge.

Seal courage notes between the halves of a walnut and fill a basket. On a day when you need a boost, pick one out and open it with a walnut cracker.

Glue cinnamon sticks to a picture frame. Add clusters of nuts and pods at each of the corners.

"Floating along on breezes, monarchs can cover eighty miles a day at speeds ranging from ten to thirty miles per hour."

-John Serrao

Autumn Leaves

One windy fall day I looked across the river and saw the leaves falling from the trees. The way the sunlight shone through the colorful leaves, there appeared to be a shower of glittery jewels. That was the beautiful inspiration for this piece. Many colors and shapes of leaves, collected on walks, are first pressed then arranged on a large canvas. Real gold leaf is added to accent the leaves. As a finishing touch the canvas is framed with sticks and made into an easel piece to be displayed as a glorious piece of nature art.

NECESSITIES

Pressed leaves
Gold leaf
Liquid podge
Paintbrush brush, 1" flat
Canvas, 16" x 20"
Sticks
Nails
Hammer
GOOP™ adhesive

CREATIVITY

1. Coat one quarter of the canvas with liquid podge. Apply bits of gold leaf by tearing off pieces of leaf and placing them on the coated canvas. After some pieces have been applied, go back to the first one and rub it with the tip of the brush. This causes the gold leaf to break into tiny pieces. Repeat for the rest of the applied gold leaf. Working in quarter sections at a time, finish applying the gold leaf to the rest of the canvas.

2. Using more liquid podge, adhere leaves in a pleasing arrangement on the canvas. Try to use the colors all over the canvas; for example, rather than all of the red in one place. Overlap some of the leaves.

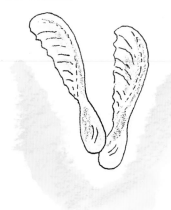

3. To make the piece have a shimmering glow, sporadically apply more gold leaf on top of the leaves. Let dry.

4. Find sticks that will extend beyond the edges of the canvas. I added 8" to the size of the sticks so they would extend 4" on the top and bottom. If you are working on a smaller canvas, use shorter sticks that are proportionate

to the canvas. Cut two sticks to size and nail in place on the sides of the canvas.

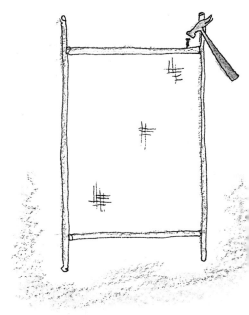

5. Measure the top and bottom of the canvas (16" in this case) and cut two sticks to size. Hammer in place on the top and bottom of the canvas.

6. If desired, make a sturdier piece by cutting two more sticks to the size of the frame legs. Glue in place.

7. For the easel stand, cut another stick 36" long. Glue stick onto the center back of the canvas, right under the top stretcher bar.

POSSIBILITIES

 Make a similar piece using summer leaves. Sprinkle it with sand while the podge is still wet.

Blow up a picture of your children or pet, mount it on the center of the canvas and surround it with an arrangement of leaves.

As a remembrance for family and friends, make tiny leaf easels featuring one beautiful leaf.

Create a mandala by placing the leaves in circles that radiate out from the center of the canvas. Use the same type of leaf in each row.

Using egg white, attach pressed autumn leaves to your windows.

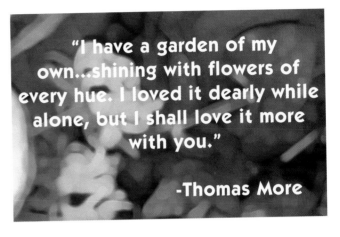

"I have a garden of my own...shining with flowers of every hue. I loved it dearly while alone, but I shall love it more with you."

-Thomas More

Scented Herb Vinegars

Versatile herbal vinegars can be used to wake up a salad or de-glaze meat juices for a wonderful gravy. Herbal vinegars can even be applied to the skin to chase away pesky mosquitoes.

NECESSITIES

Vinegar, white, apple cider or wine
Fresh herbs
Decorative glass bottles or empty wine bottles

CREATIVITY

1. Wash bottles with hot sudsy water and rinse thoroughly.

2. Place vinegar and herbs in bottle and cap or cork.

3. Allow vinegar to age. They can be used in two weeks but if aged for a year or so, the flavor is incredible. They mellow and blend beautifully. If mother forms, just remove it. Note: mother is a gelatinous starter that forms on the surface of fermenting liquids.

POSSIBILITIES

For a rich flavor add hot chili peppers, peppercorns, mustard seeds, coriander, and/or garlic cloves singly or in combination.

Some delicious combinations are:
   Salad burnet, sage and sweet marjoram in white vinegar.
   Dill, lovage, hot chili peppers and garlic cloves in white vinegar.
   Orange mint, ginger mint, cinnamon stick and allspice in cider vinegar.
   Lavender, rose-scented geranium and a lemon and lime twist in wine vinegar. (This one is nice for a facial splash.)
   French tarragon, orange mint, thyme, bay and chives in cider vinegar.
   Rosemary, opal basil and Egyptian onions in white vinegar. (This turns a beautiful pale pink color.)
   Be open to trying many combinations of the herbs you grow. The blends are a matter of personal taste, and many herbs combine beautifully. Experiment.

For delicate flavor, add heavily scented, **unsprayed** roses.

Garlic permeates the vinegar better if first put through a garlic press.

Dip the cork in wax to create an attractive seal.

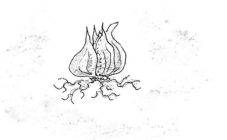

Make hand-lettered stick-on labels. Or, with a ribbon, attach name tags around the throat of the bottle.

> **"Many a visitor has departed from my garden with a handful of sweet smelling leaves carefully treasured in handkerchief or pocket."**
>
> **-Rosetta E. Clarkson**

Sleep Pillows

At the end of a long busy day it isn't always easy to get to sleep. The specially selected herbs used in these pillows have been used for years to help people drift off to dreamland. Make a second one for your travel adventures.

NECESSITIES

1/2 cup lavender flowers
1/2 cup hops
1/2 cup feverfew
1/2 cup rose petals
1/4 cup marjoram
Two 8" squares of muslin
Two 10"x12" pieces of decorative cotton cloth
Thread & needle or sewing machine

CREATIVITY

1. Mix all herbs together in a large bowl.

2. To create a pouch, sew together the two muslin pieces on three sides. Fill the pouch with herbs. Sew the end closed.

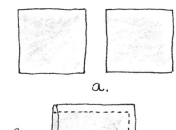

a.

b.

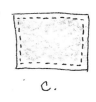

c.

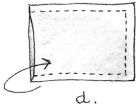

d.

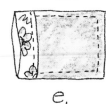

e.

f.

3. Right sides together, sew the decorative cloth pieces together. Fold the raw end approximately 2" to the inside. Sew the hem in place. Turn the cover right side out. Place the herb pouch inside the decorative, washable cover.

4. To use: Place the sleep pillow under your

regular pillow. The scent of the herbs will permeate the regular pillow giving you a restful sleep.

POSSIBILITIES

🌼 Make a sleep pillow, for a baby, using only scented rose petals.

🌼 Make a hand-size rescue pillow for your car and/or purse. Fill the bag with 3/4 cup rose petals, 1/4 cup cinnamon and one or two euca-lyptus leaves.

🌼 Find a special decorative cloth for the cover. The pillows work even better if they are personally appealing. If you are a ribbons and lace woman, go for it.

🌼 Make a mixture of herbs. Select herbs that possess the scents you love and make padded (batting) seat cushions for your kitchen chairs.

🌼 Use rose petals and scented geranium leaves to create liners for your lingerie and socks drawers.

a.

b.

c.

🌼 More sachet techniques...

a.

b.

c.

"...there comes a day...when [the lilies] are all pealing their chorus in the breeze and the whole village is drunk with a perfumed melody.

-Beverly Nichols

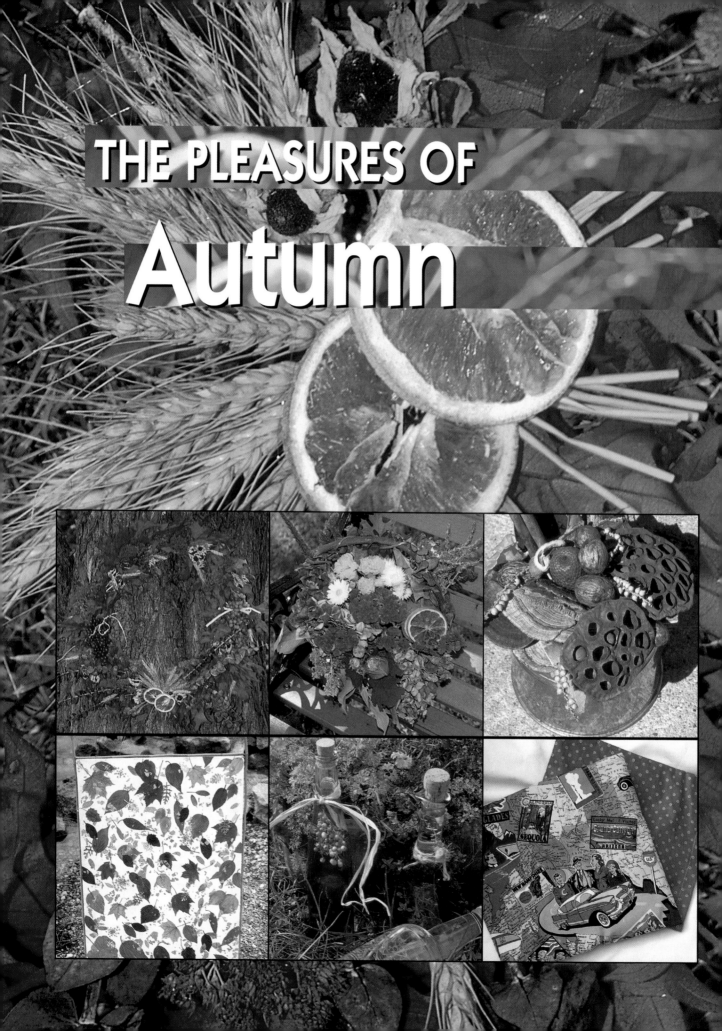

THE PLEASURES OF
Autumn

☙ Rent motorbikes and take a trip to see the astonishing fall color.

☙ Gather and press the last of the colorful flowers and leaves so you'll have materials for winter presents.

maple

☙ Place pumpkins all over your front yard and insert a sign that says "Happy Halloween - Take One."

☙ Gather leaves and preserve them yourself. Magnolia and maple leaves seem to work best. Gather branches of leaves and smash the branch end with a soft mallet. Then, insert the branches in a gallon of water to which you have added 3 1/2 cups glycerin. Place the bucket in your garage. In a few weeks you will see the leaves have absorbed the glycerin and are preserved. Wipe off the leaves and store in a dry place until ready for use.

☙ At the base of a huge pine tree in my front yard I placed three large rocks. Each autumn I collect walnuts from faraway trees and amass a pile beneath the tree. Throughout the winter the squirrels perch atop the rocks and enjoy a nut. A pleasurable winter view!

☙ This is the time to plant spring bulbs. It's easy. What a surprise you will receive in the spring. I love spring bulbs, especially tulips as they are so elegant. When the bulbs are purchased you will receive information on the depth of planting - tulips should be planted 6" deep, while scilla should be planted only a couple of inches. Basically the larger the bulb, the deeper it should be planted. Plant the bulbs in masses or around the base of large trees.

Beautiful.

Each year I plant about one hundred bulbs as not all of the previous bloomers will come back. The Appeldorns are the most hardy with a true perennial habit.

☙ While you are planting bulbs, create a lovely Christmas gift for your home or as a gift. Paperwhite narcissus are pure white star-like blossoms held on tall wispy foliage. They are tender bulbs that are easily forced in water. Select a shallow container that does not have a hole in the bottom. Add a layer of gravel, then the bulbs, planting them close together. Add more gravel just until the bulbs are covered halfway. Add water. In two weeks the flowers will bloom, producing a heady scent. Discard the bulbs when they have finished blooming.

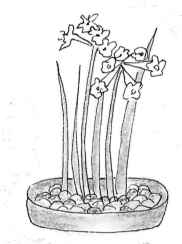

Try forcing other bulbs too. Plant them in a pot of soil. Place the pot in the refrigerator for eight weeks. Bring into the light and begin watering. Soon the bulbs will bloom. For best results pick varieties that are recommended for forcing.

Bring your houseplants, that spent the summer outside, into the house. Wash down the plants and the pots to avoid bringing in any insects.

Mulch new shrubs and trees.

Plant newly purchased perennials now. The sales are fantastic this time of year and the plants will thrive.

Cut and dry all everlastings. Until you use them, hang everlastings with twine or rubber bands from the rafters, on a hutch or on an old-fashioned clothes drying rack for display.

Harvest all of your herbs. Tie them into small bunches and hang them from a pot rack in your kitchen.

Compost your leaves. In a year or two they will decompose and become beautiful soil.

When you put away your summer outdoor equipment, be sure to move the winter tools, such as shovels and the snowblower, to a handy location for easy access.

Watch for birds in the autumn sky. They are returning to their south of the border winter homes.

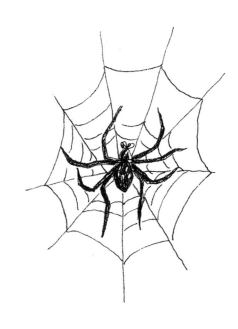

PART FOUR:
Winter

The year starts to slow down. The summer birds head south. The brilliant colors of summer turn to the deep, rich colors of autumn - golden yellow, russet, scarlet, terra cotta and mahogany. The sun is lower in the sky. It is the time to harvest the bounty of the gifts of the garden and field. It is a time of reflection.

Herb & Flower Bundles

Collect and dry as many beautiful flowers as possible. Create a huge herb and flower bundle for your door. Or, make a pair of bundles and place them on either side of your fireplace or beside a large mirror. Hang a single bundle on your front door or lay it casually on your mantle or a side table.

Salal leaves
Sweet Annie or other filler
Dried flowers
Glue gun
Florist wire
Twisted craft paper

CREATIVITY

1. Cut lengths of sweet Annie to about a 20" length. Use wire to secure the branches together to form the base of the bundle. If you are unable to find sweet Annie, use artemisia or another fragrant filler.

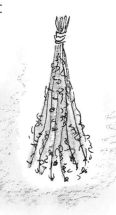

2. Glue stems of salal leaves around the edge of the bundle.

3. Begin building up the herb bundle. Starting at the bottom, glue in bunches of flowers. Overlap the bunches as you work up the b u n d l e . Sporadically add more salal leaves. When you reach the top, study the arrangement and decide if you need to add any additional flowers to fill in the design. This decorative piece should be very full, lush and colorful.

4. Add a wire loop to the top of the bundle.

5. Cut a 48" length of twisted craft paper and unfurl. Find the center of the length and tie it around the top of the bundle. Tie a bow and, if necessary, cut the ends.

POSSIBILITIES

🌿 Attach a hand-lettered banner that says "Welcome."

🌿 Make tiny bundles and hang them in a row above a chair rail.

🌿 Hang bundles from your chandelier.

🌿 Match the tie to your home decor: Twine for rustic, raffia for country, satin ribbon for Victorian or moire' ribbon for classic.

🌿 Scent the bundles with essential oil or potpourri refresher.

"Although most...gardeners...don't go into gardening with the object of improving themselves, shy people become friendly and stiff ones thaw out, and the most unsuspected talents for combining colors both indoors and out are manifested..."
-Helen Morgenthau Fox

Whistling Swan

The graceful swan, pictured here, is actually a topiary form that is covered with salal leaves. Splurge and buy yourself a dozen roses in your favorite color and immediately hang them to dry. Fill the swan with all of the roses. A sweet memory of summer past and a reminder of the new spring to come.

CREATIVITY

1. Select a topiary form. In addition to the swan, there are other shapes available : watering cans, trees, spires, hearts and more.

babies breath

2. Paint the topiary form so the metal does not show .

3. Cover the topiary form with handmade paper by tracing around the different sections. Glue in place.

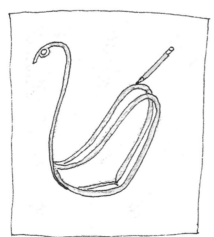

4. Add branches to any areas not covered with the paper. In the case of the swan, I used sand cherry branches to extend the tail and to create the neck /head. Glue branches in place.

5. Starting at the top of the form begin gluing leaves in place in an overlapping fashion. Maintain the shape of the topiary form and trim away any excess at the bottom and any other necessary places.

6. Use raffia to accent the piece. To distinguish the head on the swan, I wrapped the beak and tied the ends in a bow. If you are using the topiary form of the watering can, you may want to wrap the handle and tie a bow. Accent the piece you are creating in any chosen manner.

7. Insert the Oasis or Styrofoam into the topiary form. Insert the roses and add any desired filler such as artemisia or babies breath.

❦ Consider other topiary forms: Trees, butter-flies, bears, deer, pompoms and spirals.

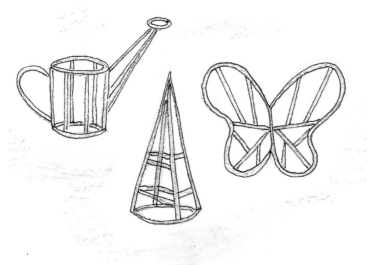

"I prefer winter and fall, when you feel the bone structure in the landscape - the loneliness of it - the dead feeling of winter. Something waits beneath It - the whole story doesn't show."
-Andrew Wyeth

❦ Use other leaves: Maple, eucalyptus, javra, etc.

❦ Decorate a bowl-shaped topiary. Insert a corresponding glass bowl and use it to serve goodies at your next party.

❦ For a rustic appearance, cover the form with abutted sticks.

❦ Fill the form with a large handful of colorful dried flowers from your garden.

Classic Pine Cone Wreath

This project takes a bit more time to create than many of the others, but it is an heirloom piece that will last for many, many years. I made this wreath over twenty-five years ago! Other pine cone wreaths were commissioned for fine dining establishments and historic buildings in my home town. You'll find the wreaths still displayed today.

Pine cones in many sizes and shapes

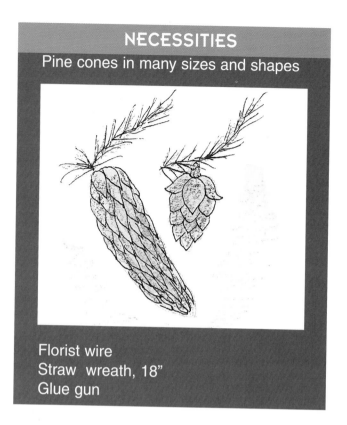

Florist wire
Straw wreath, 18"
Glue gun

CREATIVITY

1. Wire a loop around the wreath for hanging.

2. Wire the cones you wish to use by wrapping the florist's wire around the bottom of each cone.

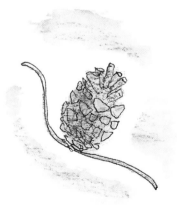

3. Begin with the largest wired cones and attach them to the wreath, then attach the mid-sized cones. Stand back from the wreath as you construct it to see if there are areas that need to be filled. Strive to obtain a design rhythm around the circular wreath form.

4. Glue in clusters of tiny cones to fill in any open spaces on the wreath.

POSSIBILITIES

🍂 Fill various-sized baskets with a single type of cone. Display them in a cluster on a table.

🍂 Attach ribbon to a single pine cone and use it for a package tie-on.

🍂 Seal the wreath in a box with one tablespoon of cinnamon. When displayed it will give off a warm, comforting scent.

🍂 To create a "lodge" look, use GOOP™ adhesive to attach pine cones to a lamp base.

🍂 Make a treat for the birds with excess cones. Apply suet to a large cone and roll it in bird seed.

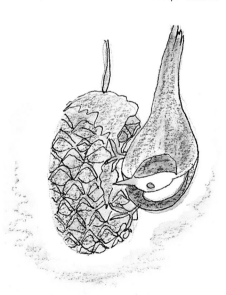

"After the hurly-burly of the final garden cleanup, we turn grate-
fully to fireside hours with books; we enjoy fragrant burning of
well-seasoned fruit woods-apple and pear, plum and cherry- and we
throw into the blaze faggots of aromatic lavender, wormwood, and
costmary that we have saved for this cold-weather pleasure, and
pine cones, too, gathered on late walks in autumn."

- Helen Van Pelt Wilson

Comfort Box

Make a special box or chest to hold personal treasures-love letters, cards from birthdays, Mother's Day and other holidays, little mementos, etc. I chose a box with brass corners and handles. I decorated it with herb and scented geranium leaves and carnation petals from my collection. This is a box filled with love. You'll visit it often.

NECESSITIES

Box or chest
Black acrylic paint
Paintbrush, 1"
Pressed leaves and flowers
Palmer liquid podge

CREATIVITY

1. Paint box. You can save cleanup late by covering any hardware with masking tape. Let the paint dry thoroughly and add a second coat if necessary.

2. Lay out the leaves on the box lid. Remove the leaves and coat the lid with liquid podge. Immediately set the leaves in place. If necessary, add additional liquid podge and let dry.

3. Create the flower design. Remove flowers and coat the area to be covered with liquid podge. Set the petals in place and let dry.

4. To seal in the leaves and flowers, coat the entire lid with another coat of liquid podge.

5. Create a design for each of the four sides of the box. Remove the leaves and flowers. Coat the box with liquid podge. Add leaves and flowers and let dry.

6. Seal entire box with a coat of liquid podge and let dry.

7. Fill the box with treasures.

POSSIBILITIES

🌿 Create a wedding box using the invitation and other wedding mementos. Remember to press some of the flowers from the bridal bouquet.

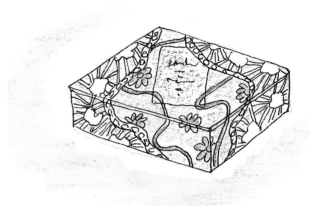

🌿 Create a travel box using maps, brochures, post cards and souvenirs.

🌿 Make a twenty-fifth wedding anniversary box. Paint it silver and add flowers in the recipients favorite colors.

🌿 Make a special box for each of your children. Use drawings, school pictures and a piece of a school paper that received an "A."

Celebrate your childhood. Make a box of things saved from your easy, carefree days.

Easy Herb Bread

Even if your time is limited you can make delicious looking and tasting herb bread. The basis of this bread is refrigerator rolls slathered in butter and herbs.

NECESSITIES

One tube refrigerator rolls
1/4 cup butter
Herbs: thyme, sweet marjoram, dill, about
    1/2 teaspoon each
Baking pan

CREATIVITY

1. Open rolls and separate.

2. Microwave butter on HIGH for one minute or until melted. Dip each roll in butter and arrange in a pan, in an overlapping fashion.

3. Sprinkle the rolls with herbs and pour on any leftover butter.

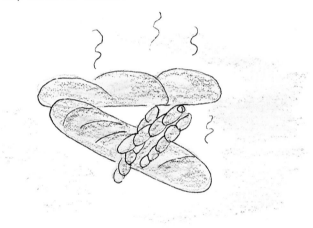

4. Bake as instructed on package.

POSSIBILITIES

🌱 Before baking, pour this herb/butter sauce over your favorite biscuits. Sprinkle with grated cheese.

🌱 Add other cooking herbs singly or in combination. Add onion flakes for a zesty treat.

🌱 Form and bake the rolls in a ring shape. Serve on a platter with a bowl of fondue.

🌱 Beware: This smells so good when it's baking that you need to be prepared; get ready for pats and hugs.

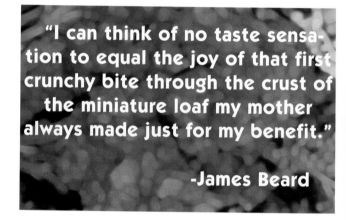

"I can think of no taste sensation to equal the joy of that first crunchy bite through the crust of the miniature loaf my mother always made just for my benefit."

-James Beard

Beauty of the Bath

Take some time to be nice to yourself. To refresh your spirits, splash on some skin freshener. Try an invigorating facial scrub. Relax with an herbal massage followed by a lingering bath. Enjoy.

Skin Freshener

For a welcome cool splash during the summer, keep this skin freshener in the refrigerator.

NECESSITIES

Vodka
Any of the mints: spearmint, peppermint, chocolate mint, eau de cologne mint or ginger mint. For a softer soothing freshener use camomile; for an exotic freshener try patchouli. Bottle with cork or twist on cap

peppermint

CREATIVITY

1. Wash and dry a bottle.

2. Fill the bottle with vodka. Add your choice of sprigs of herbs. Cap bottle.

3. Set bottle in the sunlight for one week to extract the scent and essence of the herb.

4. Strain the herbs from the vodka. Discard herbs and recap bottle. It is now ready to use.

POSSIBILITIES

🌱 Use other liquids for the base. Try unscented alcohol.

🌱 Use rosemary for a piney aroma that men find appealing.

🌱 Make small batches using different combinations of herbs. Add a drop or two of essential oil. You will come up with a signature scent that can be used as a cologne.

🌱 Place the freshener in a spray mister and use it to scent your home.

Facial Scrub

Gently scrub away dirt and grime with this soothing facial scrub. It can be used all over the body. Instructions for a special washcloth beauty bag follow the scrub.

NECESSITIES

2 cups cornmeal
2 cups grated soap
1 cup lavender flowers
1/2 cup camomile flowers
1/4 cup lovage leaves

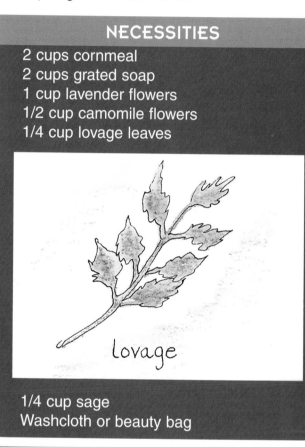

lovage

1/4 cup sage
Washcloth or beauty bag

CREATIVITY

1. Mix all scrub ingredients.

2. Place a heaping tablespoon of the scrub mixture into the center of a damp washcloth. Gather up the ends of the washcloth and gently scrub your face using a circular motion. Do not scrub the delicate skin under your eyes. Rinse your face thoroughly. Now feel your skin. Silky smooth. Beautiful. Discard the spent scrub and rinse out washcloth.

POSSIBILITIES

🌿 Use lemon verbena and lemon rind for a wake-up scrub.

🌿 Use camomile, sweet marjoram and lavender for a calming scrub.

🌿 Rose buds, lavender and orange mint make a fragrant romance scrub for a special night.

🌿 Calendula, comfrey and thyme are healing.

🌿 For a sensuous scrub, use scented geranium leaves and rose petals.

Massage Oil

According to anthropologist Ashley Montague author of Touching, our skin loves to be touched. A massage is so refreshing. Massage the whole body with this lovely oil. Slip into a soothing bath and your worries will float away.

NECESSITIES

Sweet almond oil
Essential oils
Small brown bottles
Funnel

CREATIVITY

1. Use the following chart to create a special purpose massage oil:

bay

Calming	Invigorating	Sensual
Camomile	Sage	Patchouli
Rose	Lemon	Musk
Thyme	Coconut	Ylang-ylang
Lavender	Pine	Jasmine
Bay	Eucalyptus	Sandalwood
Vetiver	Peppermint	Myrrh

2. Use the oil in a ratio of two ounces almond oil to two or three drops of essential oil. Do not use essential oils undiluted on the skin. They are too intensely concentrated and can cause health problems.

3. Use the funnel to transfer the oil to a small dark-colored bottle. Add essential oil. Cap and store in a dark cardboard.

POSSIBILITIES

🌿 The sweet almond oil is called a carrier oil in this type of formulation. Other acceptable carrier oils include: olive oil, vegetable oil, wheat germ oil, avocado oil and sesame oil.

🌿 Essential oil test: An essential oil will not stain a piece of paper. If the essential oil does stain it has already been cut with a carrier oil. This is then considered a fragrance oil. Find a different supplier.

🌿 Pier One™ carries small figural bottles that work well for this project. Check other suppliers too.

Beauty of the Bath

Luxuriate in a special herbal bath. Place the ingredients in a small drawstring bag, hang it on the faucet, then draw your bath. Lovely.

NECESSITIES

3 cups oatmeal
1 cup powdered milk
Two bars grated castile soap
1 cup lemon verbena

lemon verbena

1/2 cup lemon rind
1/2 cup orange rind

CREATIVITY

1. Combine all ingredients. Blend small batches in a blender until all ingredients are reduced in size and the mixture is thoroughly intermingled.

2. Place the mixture in a large decorative jar or fill several drawstring bags and keep them in the jar.

POSSIBILITIES

🌱 For a special bath, rub your body with oil then take the herbal bath.

🌱 For a more intense-smelling bath, add lemon essential oil.

🌱 Put on your favorite music and light a candle.

🌱 Invite a friend to share your bath.

> **"Beauty alone is useless unless it reflects the experiences of living."**
>
> **-Virginia Castleton Thomas**

THE PLEASURES OF
Winter

🍂 Take up ice skating again. The air is so crisp and clear. In addition you will be stretching your muscles. Feel supple, young and free.

🍂 After a day of skating, skiing or sledding, invite everyone back to your home for steaming mugs of hot chocolate with huge dollops of whipped cream. Shave a bit of chocolate on top. Add creme de menthe or peppermint schnapps to the adults mugs.

🍂 On warm winter days, water evergreens so they don't get winterburn.

🍂 With your camera, take pictures of the beautiful snow and ice formations and add them to your journal. Or, mount the photos in plain black frames and hang them on the wall behind a couch. Beautiful and unexpected.

🍂 Build a snowman or a snowwoman. One year before we left for Florida we built a snow mermaid about twelve feet long. She was complete with a green satin bikini and a "Gone to Florida" banner!

🍂 Take up basic knitting (knit one row, purl one row) and make scarves for everyone in your family. Use this time for meditation.

🍂 Make a quilt of simple squares with solid-colored cloth. Give visitors a Pigma™ micron permanent marker and invite them to write a message, a poem or do a drawing on one of the squares.

🍂 Don't try to remove ice from trees and shrubs. They are very brittle now and the limbs can easily be snapped off. Let the winter sun take care of the ice.

🍂 While shopping for the coming holidays, pick up something special just for you. Buy a good book, have materials on hand to start a new painting, collect things for a collage and put them in your comfort box. Don't open until December twenty-sixth or later; whenever you need a lift.

🍂 Garden catalogs begin to arrive around Christmas. Place them in a huge basket by your favorite chair. When you need a break, dig into the basket and begin to plan next years gardens. Keep pencils and a big drafting pad handy. Scissors, too, so you can add pictures to your journal.

🍂 Try a new recipe for something you've always wanted to try.

🍂 Check out the online garden bulletin boards.

Don't prune spring blooming shrubs yet. You will cut off or reduce the potential flowers. Prune in the late spring after the shrubs have bloomed. The summer bloomers, such as spirea and hydrangea, can be pruned now.

In late spring begin to awaken the trees by watering them deeply.

Treat yourself to a professional massage. Warning! This is habit-forming!

PART FIVE:
Holiday

Bring light and glamour into your life during the short, dark days of winter. This is the season of giving thanks for the gifts of the year. It is a time of celebration. Make gifts for friends, attend every party to which you are invited or throw a frolic of your own. Make merry and applaud Mother Earth. Then make time, quiet time, just for you.

Welcome Basket

Greet your guests with an elegant silver glitter weed basket adorned with deep red dried roses, wisps of fragrant eucalyptus and brilliant golden bows. Surround the basket with tiny gifts to present to visitors - homemade cookies, popcorn balls, sachets of potpourri, a crystal bead on a string or a tiny journal with an eight pack of crayons.

NECESSITIES

A large bunch of dried weeds
Silver glitter spray paint
A large white handled basket
Oasis
Moss
Twisted craft paper, gold
Florist's wire
Six dried deep red roses
Six foot-long pieces of green eucalyptus

CREATIVITY

1. Spray paint dried weeds and hang to dry.

2. Cut Oasis to fit into the bottom of the basket. Cover the Oasis with moss.

3. Insert the silver weed stems into the Oasis to create a nice full all-over arrangement.

4. Wire together three red roses and three sprigs of eucalyptus. Use wire to attach a flower bundle to either side of the basket handle.

5. Cut two pieces of twisted craft paper two feet long. Unwind the paper and tie each piece into a bow. Trim ends if necessary.

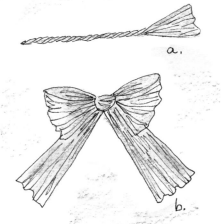

6. Insert a piece of wire through the knot on the back of the bow. Using the wire, attach the bow onto the basket handle. Repeat for the other side of the basket.

POSSIBILITIES

Spray the weeds in colors to match your mood or the season. Coordinate with complimentary roses.

Spray leaf sprigs in spectral colors. Use silvery artemisias and white roses.

When you tire of this arrangement make it into a stunning potpourri. Break the silvery weeds and leaves into small pieces, remove the rose petals from the stems and toss everything together. Add fifteen drops rose oil, five drops coconut fragrance oil and three drops musk oil for every three cups of botanicals. Seal in a glass jar for two weeks before using.

Make wish sticks to insert into the arrangement. A wish stick is a natural stick adorned with dried flowers, star tinsel and curly ribbon. Make wishes!

"In the depths of winter I finally learned that within me there lay an invincible summer."

-Albert Camus

Beaded Topiary

I love the sculptural quality of topiaries. Here the gifts of nature are adorned with the gilded accents of the holidays. One of the secrets of a successful topiary is proportion. The top ball should be larger than the container and there should be adequate space between the top ball and the container. This space should be at least twice the size of the top ball. For instance: For a 6" top ball use a middle stick at least 12" long. Make a pair to grace a mantle.

NECESSITIES

Styrofoam ball, 6"
Stick, 16" long
Oasis or Styrofoam
Moss
Invisible thread
Container
One string gold beads
Fern pins
One dried red rose
One sprig dried salal leaves
Gold twisted craft paper
Deep red twisted craft paper

CREATIVITY

1. Unfold moss and lay it on the Styrofoam ball. Wrap in place with the invisible thread. Continue adding moss and wrapping with invisible thread until the entire ball is covered.

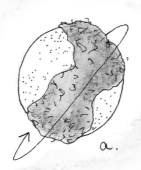

2. Insert the stick into the Styrofoam ball.

3. Place the Oasis or Styrofoam into the container and cover with moss. Insert the stick into the container.

4. On the ball, wrap the beads in place using fern pins to secure.

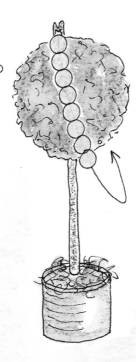

5. From each color of twisted craft paper, cut a piece long enough to fit around the container. Tie the paper around the container and make a knot in front. Twist together the papers, bring the ends to the back and tie in a knot. Stick the paper ends under the knot.

6. Insert the leaf sprig and rose into the container.

POSSIBILITIES

The paper 'butterflies,' shown in the foreground, can be used to decorate the topiary ball. Cut pieces of deep red twisted craft paper 1 1/2" long. Unfurl the paper, then hold in the center and twist to create a butterfly. Attach to the moss ball with decorative pins.

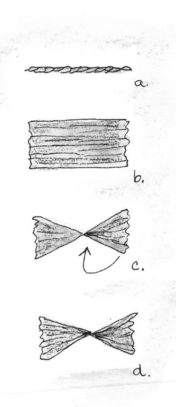

Decorate the ball with a few treasured Christmas ornaments.

Scatter the ball with scads of tiny Christmas bows in a melange of colors.

Tie beautiful French ombre ribbon around the container. Keep the ends long so they flow onto the table.

Cover the ball with cedar "curls" (better known as pet bedding).

> Collecting, dispensing, singing, there I wander with them,
> Plucking, something for tokens, tossing toward whoever is near me,
> Here, lilac, with a branch of pine,
> Here, out of my pocket, some moss which I pull'd off a live-oak in Florida as it hung trailing down,
> Here, some pinks and laurel leaves, and a handful of sage..."
>
> -Walt Whitman

Golden Red Ribbon Wreath

Humble grapevine is the perfect material for a sensuous woven wreath. Wrap the wreath in scads of gold-edged red satin ribbon. Add two white doves in flight (searching for their nest). Brighten the rest of the season, this wreath is handsome enough to be displayed beyond the holidays.

NECESSITIES

Grapevine
Florist's wire
One spool gold-edged red satin ribbon,
   1/4" wide
Three red velvet package tie-on bows, 4
   1/2" wide
Two white doves
One bag excelsior, natural color
Gold spray paint
Glue gun

CREATIVITY

1. Wind fresh grapevine into a wreath shape. Be sure to make it nice and full. This wreath was wrapped approximately twenty times. Wind the end into the wreath to hold the shape. Set aside to dry.

2. Use the florist's wire to fashion a hanging loop to the back of the wreath.

3. Spray wreath with gold paint. A little paint gives a subtle glow; a large amount of paint gives the wreath a burnished appearance. Set wreath aside to dry.

4. On the wreath, glue the topmost dove in place.

5. Wind ribbon around the wreath. Leave a long tail to hang down the front. Glue the other end of the ribbon to the doves beak. Again, leave a tail of ribbon (about 8"). Cut the end of the ribbon on an angle so the ribbon doesn't fray.

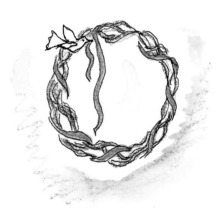

6. Glue the other dove in place.

7. Tie on the three velvet ribbons.

8. Take a 6" ball of excelsior and fashion it into a nest shape by pressing down in the center of the ball. Set the nest in place. Add glue if necessary.

POSSIBILITIES

✤ Leave the wreath natural, add realistic-looking birds and a genuine bird nest.

✤ Make a celestial wreath wrapped in star-studded tinsel.

✤ Place a Santa hat on the beribboned wreath, and tie on small Christmas-wrap packages.

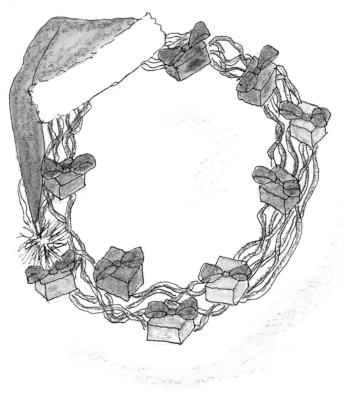

"I will no longer permit the avid and eager eye to steal away my whole attention. I will learn to enjoy more completely all the varied wonders of the earth. "

-David Grayson

✤ To create a sparkle of rainbows around the room, wrap the wreath with strands of crystal beads and hang in a window.

✤ Make tiny matching wreaths to decorate your tree.

✤ Run ribbon from wreath to wreath around the tree.

Illuminated Golden Tree

This natural tree is exciting enough to be used as the main tree for your holiday celebrations. The lights shining through the golden vines are beautiful on a dark winter night.

NECESSITIES

Grapevine
Road cone or large Styrofoam cone
One bag excelsior, natural color
One white dove
Gold spray paint
One spool gold-edged red satin ribbon,
   1/4" wide
Twenty-four red velvet package tie-on
   ribbons, 4 1/2" wide
One string of Italian lights
Masking tape
Glue gun

CREATIVITY

1. Wind a fresh grapevine wreath around the cone starting at the top. Continue down the length of the cone. Wind vine end into the wrapping to secure. Let the vine dry.

2. Wind the tree with the string of Italian lights. Try for an even placement from top to bottom.

3. Use a small piece of masking tape to cover each of the lights. Spray tree with gold paint and let dry. Remove masking tape.

masking tape

4. Take a 6" ball of excelsior and fashion it into a nest by pressing into the center of the ball. Glue the wreath in place on the top of the tree.

5. Beginning at the bottom, wrap the ribbon around the tree. Leave a length of ribbon at the top.

6. Glue the dove in place. Glue the ribbon to the beak of the dove and let a tail of ribbon hang down. To prevent fraying, trim the end of the ribbon on the diagonal.

7. Tie on the velvet ribbons.

POSSIBILITIES

🍃 This tree can be made to any size just by using a smaller or larger cone.

🍃 For the armature, try using an upside down tomato cage.

🍃 For added sparkle, hang icicle ornaments all over the tree.

🍃 Scent the tree with essential oils, hanging sachets or tiny spicy pomanders.

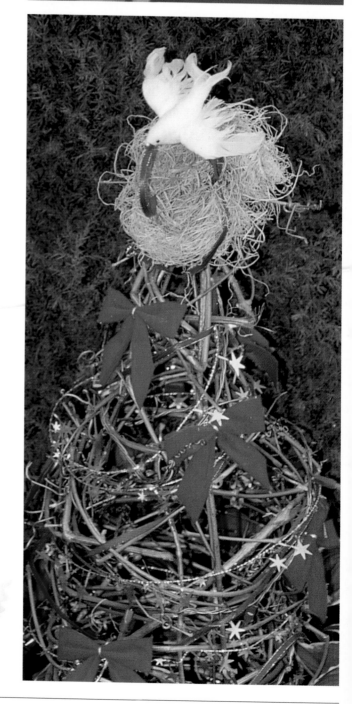

"Smell is a potent wizard that transports us across thousands of miles and all the years we have lived."

-Helen Keller

🍃 On pieces of white ribbon, write wishes with a fine line marker and tie them onto the tree. Allow them to hang down or tie into bows. Use a separate color of marker for each family member.

Holiday Dainties

These charming little nature treasures can be used as ornaments, package tie-ons and gifts for "drop-by" visitors. Or, use clusters of them to decorate your home: Around a mirror, over curtain tiebacks and around the base of a candle.

Leaf Clips

Decorate a leaf with a perfect rose and bits of decorative leaves and flowers. Clip these on a branch.

NECESSITIES

Leaf
Red rose
Bits of flowers & leaves
Paper twist craft paper, gold
Goody hair clip
Glue gun

CREATIVITY

1. Glue leaf to hair clip.

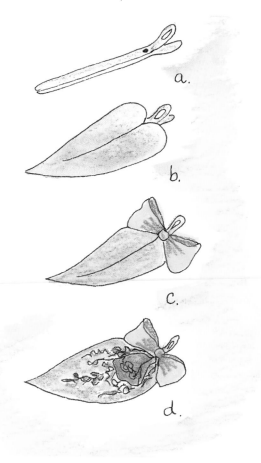

2. Take a piece of gold paper twist paper and unroll it. Fashion it into a bow by twisting it in the center. Cover the top of the hair clip by gluing the bow to the top of the leaf.

3. Glue a rose in place. Add other flowers and leaves.

Twig star

Gather a few branches and transform them into a nature star.

NECESSITIES

Clippers
Twigs
Tinsel
Glue gun
Delta™ Glitter Mist
Cord for hanging

CREATIVITY

1. For each star, cut five twigs 8" long.

2. Place the twigs in the form of a star. Carefully glue in place and allow to dry.

3. Spray the star with Delta™ Glitter Mist. This gives the star a sparkle that will be picked up by the twinkling lights, yet still creates a natural, earthy presentation. Let dry.

4. Tie tinsel on the four star points. Tie a loop of tinsel to the top point.

Earth Angel

Fashion a striking angel from nature's bounty.

NECESSITIES

Twigs
Silver paint
Rubber band
Paper twist craft paper, gold
Two salal leaves
Small Styrofoam ball
Moss
Gold cord for hanging

CREATIVITY

1. Cut twigs about 10" long. Spray with silver paint and allow to dry.

2. Wrap twigs together with a rubber band.

3. Cut a piece of gold paper twist paper. Tie paper around the twigs to cover the rubber band and create angel wings. Evenly trim ends of gold paper.

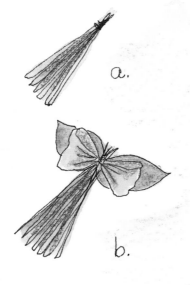

a.

b.

c.

4. Glue on leaves to create second wings behind the gold paper.

5. Glue Styrofoam ball on end of twigs. Cover the Styrofoam ball with moss.

6. Tie on gold cord for hanging.

POSSIBILITIES

🍃 Use other shapes to create twig ornaments.

🍃 Dip the shapes in glitter for a subtle glimmer.

🍃 Vary the dainties by using different colored ribbons, flowers and leaves.

🍃 For scented ornaments, coat the twigs with glue and roll them in powdered cinnamon.

" ...the best celebrations...emphasize family traditions, but also borrow colorful customs from many lands, and eras to weave a rich tapestry of prayers, parties, giving and receiving, dining and decorating."

-Adelma Grenier Simmons

Spicy Cider

The scent of cinnamon is one of the most comforting smells. Perhaps it reminds us about the love of our mother and grandmothers baking. After wrapping all of the presents, kick back and have a steaming cup of hot cider. Hot cider is also wonderful after an afternoon of ice skating.

NECESSITIES

One gallon apple cider
Allspice, 1/4 teaspoon
One long stick of cinnamon broken into 1" pieces
Nutmeg, 1/8 teaspoon
Cloves, 1/8 teaspoon
Muslin drawstring bag

CREATIVITY

1. Place all spices in a drawstring bag. Tie closed.

2. Pour the cider in a large saucepan and add spices. Simmer for five minutes.

3. Serve in mugs with cinnamon stick stirrers.

POSSIBILITIES

Add a bit of cranberry juice to the cider.

Slice peaches and serve them floating in the cider.

Cut orange strips and use them to wipe the rim of the mug. Fill mugs with cider and add the orange twist.

Serve the cider with herb cookies, lemon bread or fruit cake.

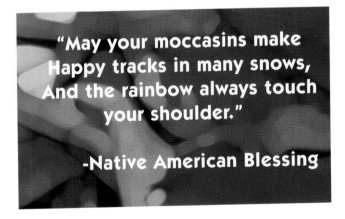

"May your moccasins make Happy tracks in many snows, And the rainbow always touch your shoulder."

-Native American Blessing

THE PLEASURES OF
Holidays

December has the shortest, darkest days of the year. Put up multitudes of clear Italian lights and use them to lengthen the day. My neighbors across the river put lights over all of their piers and leave them up until the end of February. How enchanting on a cold, dark winter night when the lights glow across the clear ice. Thank you.

Decorate your tree with a theme:

Natural Tree - Use hawthorn berries, rose hips, holly berries, tiny bouquets of red roses tied with ribbon, sprigs of babies breath and garland of gold-sprayed grape vine.

Avian Tree - Load the tree with feathered birds, tiny hand-painted birdhouses, traditional globe ornaments and a real nest at the top of the tree.

Outdoor Bird Tree - Hang decorative Indian corn, strings of popcorn and cranberries, suet balls, dried half orange cups filled with sunflower seed and raisins or sunflower heads.

Gardener's Tree - Add a watering can, a pair of garden gloves, tiny clay pots, string a hose through the branches and attach seed packets with ornament hangers.

Children's Tree - Add sports memorabilia (baseball mitt, hockey stick, football), ballet slippers, music sheets, a garden hoe or anything that pertains to their interests. Add regular ornaments for sparkle.

Ornament Ideas

🌱 Cut out images from old Christmas cards. Glue the images onto a paper lace doily. With hot glue, attach bits of evergreen and berries.

♥ Create pomander balls by inserting cloves into citrus-oranges, limes or lemons. To save your fingers, make a pilot hole with a toothpick. Attach a ribbon to the top of each ball. Roll the ball in a mixture of 1/4 cup orris root, 1/2 cup cinnamon and 1/8 cup allspice. Place on the Christmas tree to display and dry at the same time. After the holidays hang the pomanders in your closets.

♥ Go glitzy for the holidays. Wear clothing shot with gold. Apply bright red lipstick and glitter makeup. Hang your gilded wreaths. Sprinkle glitter on your side tables. Glow.

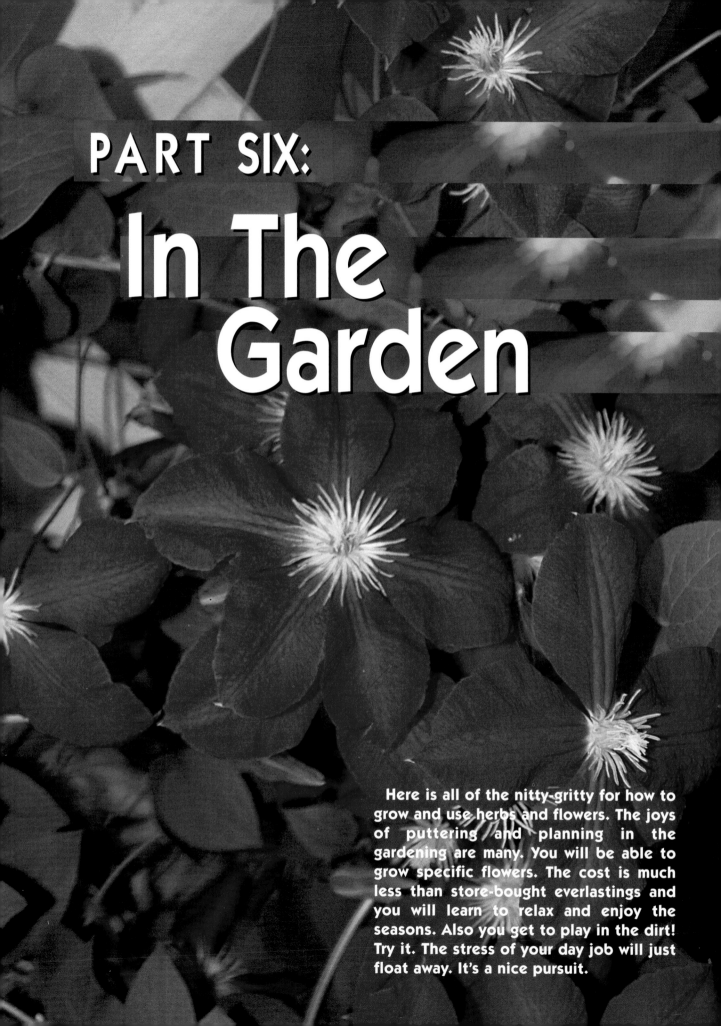

PART SIX:

In The Garden

Here is all of the nitty-gritty for how to grow and use herbs and flowers. The joys of puttering and planning in the gardening are many. You will be able to grow specific flowers. The cost is much less than store-bought everlastings and you will learn to relax and enjoy the seasons. Also you get to play in the dirt! Try it. The stress of your day job will just float away. It's a nice pursuit.

To Everything There is a Season

Every plant has a life cycle:

Annuals live for one season. They need to be replanted every spring. Petunias, dill, salvias, sweet Annie and marigolds are annuals.

Biennials complete their life cycles in two years. For instance, money plant is a biennial. The first year it sends up leaves, then dies to the ground. The second year it sends up a

money plant

flower stalk and blooms purple flowers that change to familiar silvery discs. At this time the plant has completed its life cycle and does not return.

Tender Perennials are perennials that cannot survive in colder climates. They can be planted out in the garden but must be taken indoors before the first frost. Indoors they can be grown in a sunny window or under plant lights. Scented geraniums, sweet marjoram and rosemary are tender perennials.

Perennials return for many years. Plant them once and they come back. Usually they will return, even after the harshest winters. Some perennial plants are sage, French tarragon, roses and German statice.

Each seed packet or plant stake will state the life cycle of a plant. Sometimes this is abbreviated to a letter symbol:

A = Annual
B or Bi = Biennial
TP = Tender Perennial
P or HP (hardy perennial) = Perennial

Let There Be Light

You will also find the light requirements:

FS or a white circle = Full Sun

PS or a half white, half black circle = Partial Sun

S or a black circle = Shade

Plant Zones

Use this USDA Plant Zone Chart to find your zone. Select plants that will thrive in your area. For instance, if you live in Chicago you are in Zone 5. When selecting perennial plants for your garden, make certain the label includes your zone.

Note: Most of the plants suggested in the charts are hardy from Zone 5 to 8, which covers a huge swath of our country. Many of the plants will grow in Zones 4 to 9 with some protection.

zone 3	$-40°$ to $-30°$ F.	
zone 4	$-30°$ to $-20°$ F.	
zone 5	$-20°$ to $-10°$ F	
zone 6	$-10°$ to $0°$ F	
zone 7	$0°$ to $10°$ F	
zone 8	$10°$ to $20°$ F	
zone 9	$20°$ to $30°$ F	
zone 10	$30°$ to $40°$ F	

Botanical Latin

Many plants have the same common name. This can be confusing when you go to the garden center. The special language of Greek and Latin, called botanical Latin, may seem cumbersome at first. However, this is the sure way to get exactly the plant you desire.

Linnaeus gave each plant two names, representing genus and species. The first name only, the genus, is always capitalized. For example, Lavendula dentata is the botanical name for French lavender.

The variety name is the third name (if present). Astilbe chinensis pumila is a short ground cover form of astilbe-a tall, plumed plant.

A cultivar is a plant that shows an unusual characteristic. Lavendula denata 'Linda Ligon' is a cultivar of French lavender. The grower has named this lavender after the publisher of The Herb Companion. This denotes a plant with special features - variegated leaves in this case. Often the cultivar is named in honor of a person connected to plants in some way.

A hybrid will have an "X" in the plants name. It is man-made in a greenhouse by crossing two plants. Leucanthemum x superbum Snow lady is a shasta daisy that blooms all summer long.

Some botanical Latin is very descriptive. For instance alba means white, purpura means purple, elatus means tall, cypreus means copper-like and contortus means twisted.

Get a good book on the subject and learn a few of the words. In doing so, you will become much more familiar with plant characteristics. Also, when you go to the garden center to select a small plant for the area near a walkway, you won't pick one with maximus in the name!

Selecting Seeds and Plants

To help assist you with the selection of desired seeds and plants for your garden, I've included two charts; one for herbs and another for everlastings.

chives

Herb Chart

Botanical Name	Common Name	Life Cycle	Uses and Comments
Allium schoenoprasum	Chives	P	Onion herb also available in a garlic/ onion variety (A.tuberosum).
Allium lativum	Garlic	P	The king of seasonings.
Aloe vera	Aloe	P	Skin softener, heals mild burns rapidly.

Botanical Name	Common Name	Life Cycle	Uses and Comments
Aloysia triphylla	Lemon verbena	TP	The lemoniest of the lemon herbs. Great in tea, cooking and potpourri.
Anethum graveolens	Dill	A	The pickle herb. Excellent on fish. Bolts in high heat.
Anthriscus cerefolium	Chervil	A	Anise flavored cooking herb.
Artemisia abrotanum	Southernwood	P	Used in moth preventative mixtures, dry sprigs for use in wreaths.
Agastache foeniculum	Anise hyssop	P	Nicely scented leaves for tea and potpourri.
Artemisia dracunculus	French tarragon	P	Licorice-scented herb used in vinegars and French cooking.
Chamaemelum nobile	Roman camomile	P	This is the perennial variety of the popular tea herb.
Cymbopogon citratus	Lemon grass	A	Herb used in oriental cooking. Leaves are very sharp edged.
Coriandrum sativum	Coriander Cilantro	A	The leaves of this plant are cilantro; the seeds coriander.
Foeniculum vulgare	Fennel	A	Herb seed used in cooking.
Laurus nobilis	Sweet bay	TP	Bay is a favorite in stews, soups, and bouquet garnis.
Lavendula	Lavender	P	Used in foods, moth preventatives and beauty products. Many varieties are tender perennials in northern zones. French lavender (L. dentata) is quite different in scent from the hardy English variety (L. angustifolia).
Levisticum	Lovage	P	Six feet tall! The leaves have an intense celery-like flavor. Used in soups, stews and some bath products.
Matricaria recutita	German camomile	A	This is the annual variety of the famous tea herb.
Melissa officinalis	Lemon Balm	P	Tea and baking herb.
Mentha	Mint	P	This is the huge family of refreshing mints including: spearmint, peppermint, orange mint, chocolate mint and many others. Used in cooking, crafts, teas, potpourris. Aggressive.

bergamot

Botanical Name	Common Name	Life Cycle	Uses and Comments
Monarda didyma	Bergamot	P	A tea herb. Grows wild in the many parts of the country. Attracts bees.
Nepeta cataria	Catnip	P	Tea herb and insect repellent.
Ocimum basilicum	Basil	A	One of the most used cooking herbs. Also used in potpourris and kitchen wreaths. There are many varieties of basil-lemon, cinnamon, bush, etc.
Origanum majorana	Sweet marjoram	TP	A classic cooking herb. Somewhat piney scent.
Petroselium crispum	Curly parsley	B	This is the ubiquitous plate garnish. P . neapolitanium is the more flavorful Italian parsley.
Poterium sanguisorba	Salad burnet	P	Herb with a cucumber scent. Nice in salads and herb vinegars.
Rosmarinus officinalis	Rosemary	TP	Classic cooking herb. Also use in potpourris and bath products.
Salvia officinalis	Sage	P	The turkey herb. Also used in other cooking, smudge sticks, etc. Pineapple sage (S. elegans) is a delicious scented tea herb.
Satureja montana	Savory winter	P	A bean herb. Peppery in flavor. S. hortensus (summer savory) is less intensely flavored.
Thymus	Thyme	P	Classic cooking herb. Many varieties available: caraway, coconut, lemon, etc.

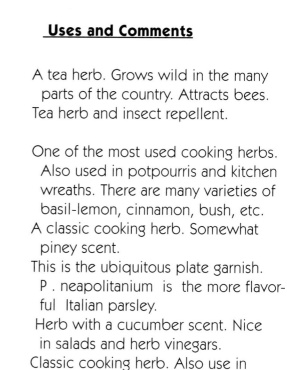

thyme

Everlastings Chart

Methods of drying:

H = hanging
Wrap a small bunch of stems together with a rubber band and hang.

D = dessicant
Commercial powders work best and can be heated and reused again and again.

AD = already dry
Allow these flowers to form a pod, then pick.

SD = screen dry
Lay flat on screen to dry.

Botanical Name	Common Name	Life Cycle	Method	Uses and Comments
Achillea	Yarrow	P	H	The yellow varieties retain their color the best.
Alchemilla vulgaris	Lady's Mantle	P	H	Beautiful chartreuse flowers.
Allium	Flowering Onion	P	AD	The seed pod is interesting.
Amaranthus	Love-lies-bleeding	A	H	An unusual flower with a Victorian appeal.
Ammobium	Winged everlasting	A	AD	Dries automatically.
Anaphalis triplinervis	Pearly everlasting	P	H	Little white buttons.
Armeria	Sea pink	P	H	White, red or pink flowers.
Artemisia	Silver King Silver Queen Mugwort	P	H	All of the artemisias listed make great filler for wreaths. SK is my favorite. Mugwort can be found growing wild. It has green leaves with a silvery backside.
Asclepias	Milkweed	P	AD	Remove the seed heads.
Astilbe	Astilbe	P	H	Spray to prevent shattering.
Celastrus	Bittersweet	P	H	Cut branches of berries just before they open.

celosia

Botanical Name	Common Name	Life Cycle	Method	Uses and Comments
Celosia	Plume flower Cockscomb	A	H	Spray the plume to prevent shattering. The cockscomb is a dense deeply embossed flower.
Chenopodium botrys	Sweet Annie	A	H	Very fragrant. Used to create herbal wreath bases. Will self-sow. Aggressive.
Chrysanthemum parthenium	Feverfew	P	H	Excellent white button flower.
Clematis	Many varieties	P	AD	Unique seedpods.
Consolida	Larkspur	A	H,D	The blue varieties are especially beautiful.
Convallaria	Lily of the Valley	P	H,D	Dries to a soft ecru.
Dahlia	Dahlia	TP	H,D	Be sure to get the drying agent between the petals.
Daucus	Queen Anne's lace	P	H	Lay flat on a screen for an open flower.

Queen Anne's lace

Delphinium	Delphinium	P	H,D	Dry loose or on stems.
Dictamnus	Gas plant	P	AD	Unique seedpods.
Dipsacus	Teasel	P	AD	Large thorny seedpod.
Echinops	Globe thistle	P	H	Not the common thistle.
Erica	Heath	A	D	Pick immature or will shatter.
Eryngium	Sea Holly	P	H	Blue flowers.
Eucalyptus	Eucalyptus	TP	H	The refreshing scent lasts.
Filipendula	Meadowsweet	P	H	Airy flowers; pick immature.
Gomphrena	Globe amaranth	A	H	Pick as soon as open.

Botanical Name	Common Name	Life Cycle	Method	Uses and Comments
Gypsophila	Baby's breath	A	H	So airy; nice filler.
Helichrysum	Strawflowe	A	AD	Wire stems, then hang.
Helipterum	Swan river	A	H	Or lay on screen to dry;wire.
Heuchera	Coralbells	P	H,D	An old-fashioned beauty.
Hydrangea	Hydrangea	P	H	Pick before fully open.
Leontopodium	Edelweiss	P	D	A flower of romance.
Liatris	Gayfeather	P	H	Huge spikes;difficult to dry.
Limonium	Sea lavende	P	H	Airy violet flowers.
	Rattail statice	A	H	Spiky flowers.
	Statice	A	H	Common statice, many colors.
Lunaria biennis	Silver dollar, Honesty	B	AD	Remove brown husks to expose pods.
Moluccella	Bells of Ireland	A	H	Unique green flower spike.
Nigella	Love-in-a-mist	A	H	Excellent pod.
Papaver	Oriental poppy	P	AD	Unique stiff-stemmed pod.

Nigella

Botanical Name	Common Name	Life Cycle	Method	Uses and Comments
Physalis	Chinese lantern	P	H	Orange pod.
Proboscidea	Unicorn plant	A	AD	Unusual pod.
Rosa	Rose	P	H,D	Hang before fully open.
Rudbeckia	Black-eyed Susan	P	H,D	Holds color well.
Salvia	Salvia	A	H	Reds turn burgundy.
Silene	Catchfly	P	AD	Small seedpods.
Spiraea	Spirea	P	AD	Large floret seedheads.
Stachys	Lamb's Ear	P	H	Silvery, furry leaves.
Tanacetum	Tansy	P	AD	Yellow buttons.
Thalictrum	Meadow Rue	P	H	Airy bouquets.
Tulipa	Tulip	P	SD	Dry petals on screen.
Verbascum	Mullein	P	H	Huge stalks of small yellow flowers.
Xeranthemum	Immortelle	A	SD	Wire and dry on screens.
Zinnia	Zinnia	A	SD	Lay flat on screens.

In addition to the flowers on the above list, try hanging grasses and ferns. Almost any flower is a possibility for screen drying for color in potpourri. Try whatever flowers and herbs you grow in your garden. There are surprises to behold.

Seed Starting

Use plastic planter flats that are available in garden centers. Or, if you don't have the space, use paper cups and place the little pots along a sunny window sill. Make holes in the bottom of each cup. Fill with soilless plant starter and plant the seed at the depth given on the seed packet. Water the cups and cover with a piece of plastic wrap held in place with a rubber band. Place on the windowsill. When the seeds sprout, remove the plastic. Water the little pots as necessary in the sink and let drain, then place back on the windowsill. If you do not have a sunny window, place the pots under grow lights and follow the same methodology.

Designing a Beautiful Garden

Here are a few tips on the different ways of how to design a luscious garden:

Design by height: In this common design the taller plants are placed in the back, medium-height plants are in the center with progressively shorter plants to the foreground. This is perfect for a background area with evergreen shrubs or manmade structures as it allows all of the plants to be celebrated. For a circular garden, work the tallest plants in the center, then work the shorter progression to the outer circumference of the circle.

Design by color: One favorite color can create a spectacular garden. A moonlight garden made up of plants in every shade and a tint of white is very dramatic. To achieve a quite sophisticated garden design, blend the plant colors as they are arranged on an artist's color wheel. Yellows move to orange-yellows, to deep oranges, to orange-reds, to bright scarlets, then deep wines and so on.

Design by plant type: Here we focus on one type of plant. A rose garden, a rock garden or an herb garden falls into this category.

Design by size: If you like miniature plants or very large plants, this style of garden is for you!

Design by style: Here one would approach the garden as an extension of the home. If you have a modern home, plant sculptural plants such as a variety of hostas. For a cottage home, a cottage-style garden of thickly massed plants, with a medley of color and form would be ideal. The idea is to match the architecture of your home to your garden.

Design by season: Spring is irresistible to those of us who garden in the north. I have one bed dedicated to the flowers of spring: bleeding hearts, tulips, daffs and many varieties of columbines. It is extraordinary in spring and fades. The spotlight is then focused on other parts of the landscape.

Design by form: Here one plants according to leaf forms. One example is a garden of round-leaved plants such as nasturtium, some circular hostas, a few round-leaved scented geraniums and eucalyptus. Think of spiky forms, serrated forms and tiny-leaved forms.

Design by scent: Scent is one of the delicious pleasures of a garden. It may be overt and loud like that from lilacs, honeysuckle and garden phlox. Or the scent may be hidden, as in many herb plants where one must rub the leaf between the fingers to aromatize. Nothing is as sensual as scent.

Design by favorites: This is a garden for the person who loves to haunt the garden centers and pick out whatever looks interesting. I'd suggest digging two plots; one in the sun and the other in shade. Plant them with your treasured plant finds. An interesting approach.

Design by use: This includes a cut-flower garden. It also includes placing edibles, such as vegetables, herbs and small fruit trees, in a potager garden.

Whatever your style, make gardening easy: If the existing soil isn't viable use raised beds, employ mulches to keep weeds down, use an underground watering system or a hose reel and quality hosing, buy yourself a pair of Wellies so the water runs right off them, get a snappy sun hat and a nicely-scented sun protectant. Have a pitcher of lemonade waiting in the refrigerator. Enjoy.

Caring for your Garden

As a general rule, newly-planted trees and shrubs should be watered deeply for five minutes each week. This allows the plants to get established so the roots multiply and go deeply into the soil. The first year, perennials also need to be tended with adequate water. After these plants acclimate to your garden they will maintain themselves. In future years they will require little excess watering. Water annuals as the weather indicates, just don't let them wilt. Enjoy annuals for their one seasons existence. Of course during an ongoing heatwave all of the plants will need more attention; during continuous wet weather do a sun dance. Use your garden wisdom.

HARVESTING

Herbs should be gathered when they are at their peak of flavor. Flowers should be harvested just before full bloom. Tie the herbs and flowers into small bunches and hang to dry. Cones and pods can be gathered as soon as they are available. Dry them in the sun before storing. If the cones dry closed, place them on a cookie sheet and place in the oven on low heat. Magically they will open.

WILD HARVEST

Mother nature provides us with useful plants. However, we must not take advantage of her generosity. Collect only what you need. Use what you've collected and be sure to leave a healthy bunch so the plant can reproduce and continue its life span. Give a silent thank-you for the plants you do take and wish the remaining plants long life. Don't collect any plants in protected areas.

WREATHMAKING BASICS

Fashioning a beautiful wreath of herbs and everlastings should conform to a style you find attractive. Some prefer tidy wreaths where the flowers are tightly attached and are of the same height; others prefer a more free expression where the plant forms extend in an erratic wild presentation. Do whichever pleases you.

To create a stunning wreath the first question you need to pose is "What kind of support will I use?" Will it be a wire form, grapevine or wild vine wreath form, straw wreath form or even a cardboard circle cut from a pizza liner?

Next, select flowers and herbs in colors that please your eye.

To attach the botanicals use fern pins, a reel of thin wire, hot glue, GOOP™ adhesive, or Tacky™ glue to miscellaneously adhere small sprigs.

It is easier to attach a hanger before the flowers are in place. Wind a piece of wire around the wreath. Secure. Stick the wire ends into the wreath form to prevent getting scratched during assembly.

Artistically you can work with a focal point, as in the autumn wreath, or do a rhythmic presentation by attaching similar groups of flowers equidistant around the circumference of the wreath. Play to discover what is attractive to you. After finding a lovely display, attach the botanicals.

To seal the arrangement from dust, spray the wreath with a hard holding hair spray.

Hang and enjoy.

CARING FOR YOUR EVERLASTINGS

Before everlasting need replacement, most last two or three years. Here are few tips to lengthen their usefulness:

🌿 Keep everlastings out of direct sunlight.

🌿 To prevent early shattering, gather the everlastings as soon possible.

🌿 Use a portable hair dryer, set on low, to dust.

🌿 Place seasonable arrangements in a box and store in a dry area.

🌿 Refresh the arrangements with hair spray.

CARING FOR CUT FLOWERS

You can extend the life of cut flowers by doing a few simple things:

🌿 Cut flowers one hour after watering so the stems are plumped with water.

🌿 Remove all leaves from the stem that will be in the vase water.

🌿 Use sharp clippers or pruners. For better water absorption, cut woody stems on an angle to expose more stem surface.

🌿 Seal ends of sap-bleeding stems by holding them over a candle.

🌿 To slow bacterial growth add a teaspoon of bleach to the vase water.

🌿 Use floral preservatives according to the package directions.

🌿 Add a teaspoon of glycerin for a preservative.

🌿 Get the flowers in water ASAP.

🌿 Arrange the flowers to delight your eye.

🌿 Remove any spent flowers and replace with fresh blooms.

🌿 For added interest, add stems of leaves to your arrangement.

COLLECTING IDEAS

Buy a three-ring binder that holds zip-lock plastic bags. Into one bag place a small pair of scissors, a glue stick, a pen, a small unlined notepad and some paper clips. As you come across ideas in magazines, newspapers, journals or other sources, cut out the idea and place it in one of the bags. Label the bags: Flower arrangements, shells, pods & nuts. Or label by room: Dining room ideas, bathroom ideas, etc. Sketch ideas on the notepad and add them to your collection. Go to the paint store and collect paint chip samples for color themes. Collect pieces of fabric to compliment your rooms. Pack the notebook with ideas and use it as inspiration for your nature art.

Planning a landscape to encourage birds includes a number of factors. They need an environment that provides areas to hide from enemies or if necessary, escape quickly. Birds need a place for making nests, courting and raising young. They also need the basics of survival; fresh food and water.

The best way to create this setting is to use a variety of deciduous and evergreen trees and shrubs; hang an assortment of bird houses with different-sized openings; add a bird bath, pond or other water source and use feeders (both hanging and platform-style). The birds will come.

TREES AND SHRUBS WITH FRUITS THE BIRDS LOVE:

Bayberry Myrics pensylvanica
Buckthorn Rhamnus spp.
Chokecherry Aronia spp.
Crabapple Malus spp
Dogwood Cornus spp.
Elder Ribes spp.
Hawthorn Crataegus spp.
Holly Ilex spp.
Honeysuckle Lonicera spp.
Mountain Ash Sorbus aucuparia
Mulberry Moraceae spp.
Oregon Grape Mahonia spp.
Russian Olive Elaegnus angustifolia
Salal Gaultheria shallon
Service Berry Amelanchier
Stone Fruits Prunus spp.
Viburnum Viburnum spp.
Winterberry Ilex verticillata

If you allow the plants to stand through the winter, birds will eat many tree fruits and many flower seeds. It will not give a tidy look, but oh, the birds and other wild creatures you will attract.

PLANTS FOR ATTRACTING HUMMINGBIRDS

The rufous and the ruby-throated hummingbirds are the most common of the twenty different species found in the United States. The tiny bird weighs less than a nickel

and flits from flower to flower quite rapidly. Place a feeder or some of the following plants close to windows or an outdoor sitting area. Enjoy the beauty of hummingbirds.

If you see a tiny cup-shaped nest woven of plants, mosses and lichens wrapped with spider webs, you have found a hummers home. The birds lay two tiny bead-sized white eggs which are incubated for two to three weeks. After twenty-five days the nestlings are on their own.

This listing encompasses all types of plants including trees, shrubs, vines and flowers. Remember that the hummers prefer flowers in hot colors such as red, orange and pink.

Flowers
P-Bergamot Monarda didyma
P-Bleeding Heart Dicentra spp.
P-Blazing Star Liatris spp.
P- Cardinal flower Lobelia cardinalis
P-Columbine Aquilegia spp.
P-Coral Bells Heuchera spp.
A- Dahlia Dahlia spp
P- Delphinium Delphinium spp.
P-Foxglove Digitalis spp.
A-Fuchsia Fuchsia spp.

P-Garden Phlox Phlox spp.
P-Gladiolus Gladiolus spp.
P-Lungwort   Artemisia vulgaris
A- Petunias Petunia x hybrids
P-Penstemon Penstemon spp.
P-Poker plant Kniphofia uvaria
A- Scarlet Sage Salvia splendens
A- Snapdragon Antirrhinum majus

Vines
P- Clematis Clematis spp.
P- Honeysuckle Lonicera spp.
A- Morning Glory Ipomea spp.
A- Scarlet Runner Bean Phaseolus coccineus
P- Trumpet Flower Campsis radicans

Shrubs
Azalea Rhododendron spp.
Beauty Bush Kolkwitzia amabilis
Butterfly Bush Buddleia spp.
Elderberry Sambucus spp.
Flowering Currant Ribes sanguineum
Flowering Quince Chaenomeles spp.
Honeysuckle Lonicera tartarica
Rose of Sharon Hibiscus syriacus
Weigla Weigla florida

Trees
Black Locust Robina pseudo-acacia
Crabapple Malus spp.
Hawthorn Crafaegus spp.
Madrone Arbutus menziesii
Red Horse chestnut Aesculus camea
Silk Tree Albizzia julibrissin
Strawberry Tree Arbutus unedo

PLANTS FOR ATTRACTING BUTTERFLIES

Butterflies need heat to fly and they prefer feeding on the nectar of sun-loving plants. Situate your butterfly garden on the south or southeast side of your house. Add a large flat rock so the butterflies can bask in the sun. Add a sun puddle by sinking a shallow pail, in sand, flat in the garden. Keep it moist so the butterflies can absorb the salts and vital trace minerals. There is nothing quite so beautiful as a yard full of flitting butterflies. They are a nice symbol of both fragility and strength.

The following list includes both nectar and larval plants. Those with an asterisk are particularly good for attracting butterflies. Choose a mixture for your garden that will flower throughout the growing season.

FLOWERS, HERBS, GROUND COVERS AND SHRUBS

Ageratum Ageratum spp.
*Allium Allium spp.
Alyssum Lobularia maritima
*Aster Aster spp.
*Bee Balm Monarda didyma
*Black-eyed Susan Rudbeckia spp.
*Butterfly bush Buddleia davidii

*Butterfly weed Asclepias tuberosa
*Choke cherry Prunus virginiana
*Coreopsis Coreopsis spp.
Coriander Coriandrum sativum
Cosmos Cosmos spp.
Dahlia Dahlia spp.
*Echinachea Echinachea purpea
*Fennel Foeniculum vulgare
*Gayfeather Liatris spp.
Germander Teucrium chamaedrys
Heliotrope Valeriana officinalis
*Honeysuckle vine Lonicera spp.
*Hyssop Hyssopus officinalis
Kinnikinnik Arctostaphylos uva-ursi
Lavender Lavendula spp.
*Lily Lilium spp.
Marigold Tagetes spp.
*Mint Mentha spp.
*Parsley Petroselinium crispum
* Phlox Phlox paniculata
* Pin Cushion flower Scabiosa spp.

*Red-twig dogwood Osier spp.
*Red Valerian Centranthus ruber
Rock cress Aubrieta deltoidea
* Stonecrop Sedum spp.
Thyme Thymus spp.

As a special nocturnal pursuit you may want to try to attract moths. Although they are less colorful, they are interesting. Differentiate a moth from a butterfly by examining the wings. A moth has wings that fold back on a broad body and they make cocoons rather than a chrysalis. Grab a flashlight and take a moth walk.

Attractants

Evening Primrose Cenothera spp.
Four O'clocks Mirabilis spp.
Iris Iris spp.
Lilac Syringa spp.
Mock orange Philadelphus spp.
Petunia Petunia x hybrids
* Phlox Phlox spp.
* Yucca Yucca spp.

A MOONLIGHT GARDEN

An attractive moonlight garden is made up of white-flowered and silvery-leaved plants that intensify the last light of the day. Many of these flowers are heavily scented which adds to your pleasure. Each evening sit in your moonlight garden and allow the busyness of the day to melt away. Drink in the fragrance and do a relaxation meditation.

Shrubs/trees and vines

Clematis	Clematis spp.	'Sweet Autumn' is a riot of small flowers; 'Henyri' is a large-flowering variety. P.
Cotoneaster	Cotoneaster multiflorus	Very hardy shrub with mid-spring flowering.
Mock orange	Philadelphus spp.	This old-fashioned shrub is hardy to Zone 5. The white flowers have a light lily-like fragrance.
Moonflower	Ipomoea alba	Huge white flowers that unfurl in the evening. This annual needs to be started indoors in cold climates. A.
Spirea 'Snowmound'	Spirea nipponica	A beautiful small shrub that looks as if snow has fallen on it.
Stewartia pseudocamellia		Summer bloomer with large soft white flowers.
White forsythia	Abeliophyllum distichum	This unusual variety blooms white fragrant flowers.

| White lilac | *Syringa* spp. | Perennial favorite spring bloomer. |
| Wisteria | *Wisteria alba* | Huge vine with pendulant flower clusters. P. |

Perennials and annual flowers

Artemisia	*Artemisia* spp.	'Silver King,' 'Silver Queen,' and 'Silver mound' all have interesting silver foliage. P.
Bleeding heart	*Dicentra alba*	A pure white version of distinct hearts on arching stems. Very romantic. P.
Dusty Miller	*Senecio cineraria*	A common silver-leaved plant. A.
Echinacea	*E.* 'white swan'	A white variety of a long-blooming favorite. P.
Geranium		Pick white varieties like 'Orbit white' with large heads. A.

Lamb's ear	*Stachys labiatae*	Fuzzy silver gray leaves. P.
Lily of the Valley	*Convallaria majalis*	Charming, fragrant spring bell-shaped flowers. P
Lisianthus		Mermaid white is a long-bloomer.
Nicotiana	*N. x sanderae*	Domino hybrids of pure white are a good choice. A.
Statice	*Limonium bellidifolium*	Sprays of white flowers. P.
White petunias	*P. x hybrida*	Pick white singles or doubles. A.
White stock	*Matthiola incana*	Spikes of fragrant white flowers. A.

When selecting plants for your moonlight garden don't forget white tulips, white roses and any other charming white flowering plants you find at the garden center. Add whatever appeals to you.

FRAGRANT ROSES

This quick guide highlights the most fragrant roses. They are excellent for drying, potpourri and simply nuzzling to the nose. Those starred are the easiest to grow.

* Abraham Darby Apricot Old English rose
* Baronne Prevost Pink Antique rose
* English Garden Apricot Old English rose
Fragrant Memory Pink hybrid tea rose
French Perfume Yellow-rose bicolor hybrid tea rose
* Gertrude Jekyll Pink Old English rose
Gina Lollobrigida Yellow hybrid tea
*Graham Thomas Yellow Old English rose
Jardins de Bagatelle Pink Grandiflora rose
* King's Ransom Yellow hybrid tea rose
Love Potion Dark lavender floribunda rose
* Madame Plantier White Old English rose
Oklahoma   Red hybrid tea rose
*Old Blush Pink antique rose
* Othello   Red Old English rose
* Paul Neyron Pink antique rose
* Penelope Pink antique rose
Red Masterpiece Hybrid tea rose
Sheer Bliss Pink hybrid tea rose
Sun Goddess Yellow hybrid tea
Sun Sprite Yellow floribunda rose
* The Pilgrim Yellow Old English rose
* Wenlock   Red Old English rose

"Working in the garden...gives me a profound feeling of inner peace. Nothing here is in a hurry. There is no rush toward accomplishment, no blowing of trumpets. Here is the great mystery of life and growth. Everything is changing, growing, aiming at something, but silently, unboastfully, taking its time."

-Ruth Stout

Resources

Avatar's World

106 East Hurd Road
Edgerton, WI 53534
608/844-4730
Dried flowers

Bunch of Bloomers

3187 Keller Road
St. Thomas, PA 17252
717/369-4951
Dried and pressed flowers

Creative Craft House

P.O.Box 2567
Bullhead City, AZ 86430
PHONE?
Seashells, Sea Horses, Dried Flowers
and Pods

The Essential Oil Company

P.O.Box 206
Lake Oswego, OR 97034
503/697-5992
Essential oils and incense supplies

The Flowery Branch

P.O.Box 1330
Flowery Branch, GA 30542
PHONE?
Seeds for everlastings and herbs

Greenfield Herb Garden

P.O.Box 9
Shipshewana, IN 46565
219/768-7110
Herb & flower books and supplies

Goodwin Creek Gardens

P.O.Box 83
Williams, OR 97544
541/846-7357
Herb and dried flower plants
and seeds

Meadows Direct

13805 Hwy. 36
Onslow, IA 52321
319/485-2723
Wholesale source of dried flowers
and foliage

San Francisco Herb Co.

250 14th Street
San Francisco, CA 94103
415/861-7174
Bulk herbs and essential oils

Suggested Reading & Books Cited

"Yes, they are tiny growing things and they might be crocuses or snowdrops or daffodils," she whispered. She bent very close to them and sniffed the fresh scent of the damp earth."
-Frances Hodgson Burnett

The Wise Garden Encyclopedia, Harper Collins, New York, 1990

The Herb Society of Cincinnati (compiler) Herbs From Cultivation to Cooking Pelican Publishing Co., Gretna, Louisana, 1979

Xerces Society (compiler) Butterfly Gardening, Sierra Club Books, San Francisco, 1990

Ackerman, Diane A Natural History of the Senses, Random House, New York, 1990

Beard, James The James Beard Cookbook Dell Publishing, New York, 1959

Beston, Henry Herbs and the Earth, David R. Godine Publishers, Boston, 1990

Buckman, Dian Dincin Feed Your Face Duckworth, London, 1973

Burnett, Frances Hodgson The Secret Garden, Frederick A. Stokes, New York, 1911

Burroughs, John The Gospel of Nature, Applewood, Golden, CO, 1990

Camus, Albert Notebooks 1942-1951, Marlowe & Co. New York, 1994

Clarkson, Rosetta E. Magic Gardens, Macmillan, New York, 1939

Clarkson, Rosetta E. Green Enchantment: The Magic Spell of Gardens, Macmillan, New York, 1940

Clarkson, Rosetta E. Herbs: Their Culture and Uses Macmillan, New York, 1940

Dana, Mrs. William Starr (Frances Theodora Parsons) How to Know the Wild Flowers Houghton Mifflin, Boston, 1989

Dillard, Annie Pilgrim at Tinker Creek, Harper & Row, New York, 1974

Druse, Ken The Natural Garden, Clarkson N. Potter, Inc.,New York, 1989

Emerson, Ralph Waldo The Selected Writings of Ralph Waldo Emerson, Random House, New York, 1950

Edison, Thomas Diary & Sundry Observations of Thomas Alva Edison, Dogobert D. Runes (ed.), Greenwood, Springfield, IL 1968

Fox, Helen Morganthau The Years in my Herb Garden, Macmillan, New York, 1953

Genders, Roy Perfume in the Garden,The Garden Book Club, London, n.d.

Grayson, David A Country Journal Random House, New York, 1989

Grenfell, Diana and Roger Grounds, The White Garden, Trafalgar Square Publishing, North Pomfret, VT, 1991

Gunn, Penja Lost Gardens of Gertrude Jekyll, Macmillan, New York, 1991

Harper, Pamela J. Designing with Perennials Macmillan, New York1991

Jekyll, Gertrude A Gardener's Testament, Scribner's, New York, 1937

Jones, Pamela Just Weeds, Prentice Hall Press, New York, 1991

Kakuzo, Okakura The Book of Tea, Charles E. Tuttle Company, Tokyo, 1956

Keats, John Complete Works of John Keats (5 vol) Forman H. Bouxton, ed. AMS Press, New York, n.d.

Keller, Helen Story of my Life, Doubleday,Garden City, NJ, 1954

Kennedy, John F. A Nation of Immigrants Harper Collins, New York, 1986

Lacy, Allan The Garden in Autumn The Atlantic Monthly, New York, 1990

Lacy, Stephen The Startling Jungle, David. R. Godine Publishers, Boston, 1991

Lawrence, D.H. The Complete Short Stories, Viking Press, New York, 1950

Lawrence, Elizabeth Through the Garden Gate, The University of North Carolina Press, Chapel Hill, NC, 1990

Linn, Debra Sacred Space, Random House, New York, 1995

Loewer, Peter The Evening Garden, Macmillan, New York, 1993

Lubbock, John The Uses of Life, Ayer, North Stratford, NH, 1977

Martin. Tovah The Essence of Paradise, Little, Brown and Company, New York, 1991

Montagu, Ashley Touching, Columbia University Press, New York, 1971

More, Thomas The Correspondence of Sir Thomas More, Elizabeth Roges (ed.) Ayer, North Stratford, NH, 1977

Morris, Edwin T Fragrance, Scribner's, New York, 1986

Neal, Bill Gardener's Latin, Algonquin Books of Chapel Hill, Chapel Hill, NC,1992

Nichols, Beverly Sunlight on the Lawn, E.P.Dutton, New York, 1957

Niethammer, Carolyn American Indian Food and Lore, Collier Books, New York, 1974

Nin, Anais Ladders to Fire, Swallow Press Chicago, 1959

Orbach, Barbara Scented Room, Clarkson Potter, New York, 1986

Peterson, Roger Tory Eastern Birds, Houghton Mifflin, Boston, 1980

Rainer, Tristine The New Diary Jeremy P. Tarcher, Los Angeles, 1978

Rose, Jeanne, Jeanne Rose's Herbal Body Book, Grosset & Dunlap, New York, 1976

Serrao, John Nature's Events, Stackpole Books, Harrisburg, PA, 1992

Simmons, Adelma Grenier Herbs Through the Seasons at Caprilands, Rodale Press, Emmaus, PA, 1987

Stoddard, Alexandra <u>Creating a Beautiful Home</u>, William Morrow, New York, 1992

Stout, Ruth <u>How to Have a Green Thumb Without an Aching Back</u>, Cornerstone Library Publications, New York, 1955

Thomas, Christopher <u>In Search of Lost Roses</u>, Summit Books, New York, 1989

Thomas, Virginia Castleton <u>My Secrets of Natural Beauty</u> Keats Publishing, New Canaan, CT , 1972

Thoreau, Henry David <u>Walden and Other Writings</u>, Bantam, New York, n.d.

Tisserand, Robert <u>The Art of Aromatherapy</u>, Inner Traditions, New York, 1977

Whitman, Walt <u>Leaves of Grass</u>, Modern Library, New York, 1959

Wilder, Louise Beebe <u>The Fragrant Path</u>, Macmillan, New York, 1932

Wilson, Helen Van Pelt and Leonie Bell, <u>The Fragrant Year</u>, M. Barrows & Co., New York, 1967

Wyeth, Andrew <u>Andrew Wyeth:Autobiography</u>, Bulfinch Press, New York. 1995

About the Author

Linda Fry Kenzle is the author of nine other books including Scented Geraniums: Enchanted Plants for Today's Garden, and Vines For America. She also has written for numerous garden publications.

Linda is a landscape designer, wildflorist, photographer, illustrator, visual artist, jewelry designer and poet. She tends her own garden paradise along the shores of the Fox River in northern Illinois.

You can reach her at lkenzle@concentric.net.

GATHERING IDEAS

GATHERING IDEAS

GATHERING IDEAS

GATHERING IDEAS

GATHERING IDEAS

GATHERING IDEAS

GATHERING IDEAS

GATHERING IDEAS

Arts&Crafts

America's Premiere Crafts Magazine is offering a special FREE Trial issue.

Try the easy "how-to" projects and craft ideas.

If you like the first issue, you can receive **5 more for only $14.99.** *A savings of more than 30% off the cover price.*

Act now to receive your **FREE** trial issue of *Arts & Crafts magazine* from Krause Publications, publishers of quality magazines and books since 1952.

Send this card in TODAY to receive your FREE trial issue of Arts&Crafts

- Outstanding, full-color photography to guide you
- Projects for crafters of all skill levels, beginner to expert
- Complete step-by-step directions on all the latest craft projects
- Reader-friendly "how-to" makes creating beautiful crafts fun and easy

YES! *Send a FREE trial issue of ARTS & CRAFTS.* If I like *Arts & Crafts* magazine, I'll pay the new subscriber invoice of **$14.99**, entitling me to a one-year subscription of 6 issues (5 additional). If not, I'll just mark "cancel" on the invoice, return it and owe nothing. The FREE issue is mine to keep.

Name _____

Address_____

City_____State/Zip _____

Telephone_____

ABARY2

Pass along this special FREE trial issue offer to a friend.

Arts&Crafts

- Complete step-by-step directions on all the latest craft projects
- Reader-friendly "how-to" makes creating beautiful crafts fun and easy
- Outstanding, full-color photography to guide you
- Projects for crafters of all skill levels, beginner to expert

YES! *Send a FREE trial issue of ARTS & CRAFTS.* If I like Arts & Crafts magazine, I'll pay the new subscriber invoice of $14.99, entitling me to a one-year subscription of 6 issues (5 additional). If not, I'll just mark "cancel" on the invoice, return it and owe nothing. The FREE issue is mine to keep.

Name _____

Address_____

City_____State/Zip _____

Telephone_____

ABARY2

Pass along this special FREE trial issue offer to another friend.

Arts&Crafts

- Outstanding, full-color photography to guide you
- Projects for crafters of all skill levels, beginner to expert
- Complete step-by-step directions on all the latest craft projects
- Reader-friendly "how-to" makes creating beautiful crafts fun and easy

YES! *Send a FREE trial issue of ARTS & CRAFTS.* If I like *Arts & Crafts* magazine, I'll pay the new subscriber invoice of $14.99, entitling me to a one-year subscription of 6 issues (5 additional). If not, I'll just mark "cancel" on the invoice, return it and owe nothing. The FREE issue is mine to keep.

Name _____

Address_____

City_____State/Zip _____

Telephone_____

ABARY2